THE COUNSELING PROCESS

Fourth Edition

Lewis E. Patterson
Elizabeth Reynolds Welfel
Cleveland State University

Brooks/Cole Publishing Company
Pacific Grove, California

I(T)P ™
The trademark ITP is used under license.

A CLAIREMONT BOOK

Brooks/Cole Publishing Company
A Division of Wadsworth, Inc.

Printed in the United States of America

10 9 8 7 6 5 4 3 2 1

Library of Congress Cataloging-in-Publication Data

Patterson, Lewis E.
The counseling process / Lewis E. Patterson, Elizabeth Reynolds Welfel. — 4th ed.
p. cm.
Includes bibliographical references and index.
ISBN 0-534-23268-X :
1. Counseling. I. Welfel, Elizabeth Reynolds, [date].
II. Title.
BF637.C6P325 1993
158'.3—dc20 93-37295
CIP

Sponsoring Editor: *Claire Verduin*
Editorial Associate: *Gay C. Bond*
Production Coordinator: *Fiorella Ljunggren*
Production: *Lifland et al., Bookmakers*
Manuscript Editor: *Jane Hoover*
Permissions Editor: *May Clark*
Interior Design: *Quica Ostrander*
Cover Design: *Vernon T. Boes*
Cover Illustration: *Lisa Thompson*
Interior Illustration: *Gail Magin*
Typesetting: *Bookends Typesetting*
Printing and Binding: *Malloy Lithographing, Inc.*

TO

Janice, my companion in all things
L.E.P.

Brandon and Fred, who bring me joy
E.R.W.

CONTENTS

PREFACE

This fourth edition of *The Counseling Process* is intended for use in introductory courses in counseling at the graduate or advanced undergraduate level. We expect this edition to be used as the earlier ones have been—as a single text in a counseling laboratory or pre-practicum course or as one of two texts in a course on counseling theory and process. The book is also suitable for use in the training of social workers, rehabilitation counselors, substance abuse counselors, paraprofessionals, volunteers who staff hotlines and others who work in human service and mental health settings. The third edition was used in schools of theology as a guide to pastoral relationship building and counseling.

As in previous editions of *The Counseling Process,* our purpose is to help readers develop an understanding of the generic principles of effective helping, regardless of the theoretical orientation of the counselor. We define counseling as a process with three major stages and delineate the knowledge, attitudes, and skills we see as essential for effective helping at each stage. Our view is that counselor and client work together as partners to understand the issues the client faces, illuminate their causes, and then establish more effective actions. New to the fourth edition are a greater emphasis on diagnosis and goal setting and a more detailed description of how the generic process of helping ought to be adapted to meet the needs of several kinds of clients—those of diverse cultural backgrounds, women, children and adolescents, clients in crisis, and reluctant clients. This edition also includes more discussion of issues facing counselors who work in community mental health agencies and other mental health settings.

The broad outline of the counseling process described in the third edition has been retained, but the material has been reorganized and expanded. The book is now organized into three major parts: Fundamentals of the Counseling Process, Adapting the Counseling Process to Diverse Clients, and Theoretical and Professional Issues. There are three entirely new chapters in the book. The first focuses on counseling clients in crisis, delineating the ways in which the three-stage model needs to be adapted for these cases. All counselors encounter clients in crisis at times, and this chapter provides the essential knowledge for intervening in such circumstances. The second new chapter attends to the counseling skills necessary to work effectively with female clients and those of culturally diverse backgrounds. This new chapter integrates the growing body of literature on these topics and provides specific guidelines for helping counselors develop the sensitivity and skills they need to be effective with these kinds of clients. The third new chapter elaborates on ethical standards and issues in counseling, an area that has received great attention in the professional literature since the last edition of this book was published.

All of the other chapters have been substantially rewritten and updated. Commitment to action and termination are now covered in one chapter, and the material on the relationship of goal setting to action planning has been greatly expanded. The chapter on diagnosis has undergone the most substantial revision and now includes a six-component model of effective diagnosis. Also, the material on counseling theories has been expanded to give greater attention to cognitive models of counseling. The chapter on research in counseling has been dropped from this edition to make room for the additional practice-oriented material.

The book employs three learning aids to facilitate students' understanding of the material. First, case studies accompany most chapters. These cases include questions that encourage the reader to place himself or herself in the role of helper and to consider client material from a counselor's point of view. The cases in this edition are written to reflect the cultural diversity in American society. Second, the book includes several exercises that may be completed individually or with partners to assist in the internalization of learning and give practice in using specific counseling skills. Finally, Appendix C provides a format for counselors who wish to evaluate their effectiveness after a counseling session. We have found this format particularly useful in practicum and internship supervision.

We have written the book in a personal style geared to the student interested in becoming an effective helper, and at the same time accessible to the layperson interested in understanding what professional counseling is all about. After using the text in an introductory course, the student should have sufficient knowledge to begin contact with clients under careful supervision. The text is also a valuable resource for students in internships who wish to review principles of effective counseling as they embark on that stage of their training.

We wish to acknowledge the contributions of our reviewers, whose careful reading of the manuscript helped us refine our thinking and writing. They are Arthur J. Clark of St. Lawrence University, Sheldon Eisenberg, Stephen S. Feit of Idaho State University, and Brenda Freeman of the University of Wyoming. The support and encouragement of Claire Verduin and Fiorella Ljunggren of Brooks/Cole have been especially valuable, as has the assistance of Jane Hoover of Lifland et al., Bookmakers.

We also wish to thank Toni Foster for her masterful word processing and unflagging good spirit throughout the writing process. We acknowledge the help of Nancy Waina and James Toman, our graduate assistants, for careful library research and indexing. We owe a debt of gratitude to Dr. Daniel J. Delaney for his germinal ideas in creating the first edition of *The Counseling Process* with our former colleague, Sheldon Eisenberg. Dr. Eisenberg's contributions to the first three editions and the stimulus of his thinking about counseling have influenced the form and substance of the fourth edition. Finally, we wish to thank our families for their support and patience during the long months we spent at the computer writing and rewriting this book.

Lewis E. Patterson
Elizabeth Reynolds Welfel

PART ONE

FUNDAMENTALS OF THE COUNSELING PROCESS

Chapter 1

Some Perspectives on Effective Helping

People seek the services of professional helpers—counselors, social workers, psychologists, and psychiatrists—when their capacities for responding to the demands of life are strained, when desired growth seems unattainable, when important decisions elude resolution, and when natural support systems are unavailable or insufficient. Sometimes the person in need of help is urged or required to seek counseling by a third party—spouse, parent, employer, teacher, or judge—who believes that the individual is failing to manage some important aspect of life effectively.

The purpose of counseling, broadly conceived, is to empower the client to cope with life situations, to reduce emotional stress, to engage in growth-producing activity, and to make effective decisions. As a result of counseling, the client increases his or her control over present adversity and present and future opportunity.

THEORETICAL FOUNDATIONS OF THIS BOOK

Persons of any age, in any walk of life, and with almost any kind of problem can be helped to gain power over the adversities and opportunities of their lives. Counseling to achieve client empowerment is viewed in this text as a generic process that includes essentially the same elements whether performed in a community counseling clinic, a rehabilitation center, a school, a hospital, or any other facility. Certain methods may result in more effective work with certain clients (see Part Two for discussion of counseling with children, women, minorities, reluctant clients, and clients in crisis), but the *basic* structure of the helping process is the same.

The power to live effectively is enhanced by knowledge of two kinds: understandings about self and understandings about conditions of the environment. Aspects of self include capacities, knowledge, emotions, values, needs, interests, and ways of construing self and others. Conditions of the environment include, among other factors, interpersonal contacts with family, friends, and work or school associates; social and economic factors associated with where one lives and how one earns a living; and cultural heritage. Because we believe that a client's control over his or her destiny is enhanced by increased understanding of self in one's environment, the generic model we present in this book must be regarded as an insight approach to counseling. We believe that a client who develops improved insights about his or her needs, desires, and capacities in relation to the opportunities and challenges afforded by his or her particular environment will be empowered to live more effectively.

The model of counseling presented here focuses on both cognitive and affective issues. We recognize that the ability to solve cognitive problems is enhanced by a positive affective state and that progress in solving cognitive problems is likely to make the client feel better about himself or herself. Improvement in either the cognitive or the affective domain enhances the ability to behave more effectively.

In our view, no one personality or learning theory accounts for all of human experience, and no one counseling approach embodies the "whole truth" about the helping process. Our generic model for counseling integrates ideas from most contemporary approaches to counseling, and subsequent chapters of this book acknowledge the special contributions of the creators of those approaches.

The three-stage model of helping is introduced in Chapter 2 and elaborated upon in subsequent chapters. The first stage emphasizes the quality of the counseling relationship and is based heavily on the work of Carl Rogers (1942, 1951, 1961, 1980, 1986) and other authors who subsequently devised methods of teaching and evaluating relationship skills (Carkhuff, 1969a, 1969b; Egan, 1990). The second stage of counseling focuses on helping clients gain deeper understandings of self. Here we rely on psychoanalytic concepts of ego function and defense of the ego, as well as unconscious motivation, to provide a basis for a counselor's understanding of a client's motivation. The counseling techniques suggested for promoting in-depth understandings of self emerge from a broad literature on counseling but have special relevance to Freudian procedures (Alexander, 1963; Strachey, 1964) and Gestalt procedures (Perls, 1969; Polster & Polster, 1973). The third stage of counseling involves goal setting and action planning. We have relied on the work of trait-factor counselors (Williamson, 1939, 1950, 1965) and cognitive-behavioral counselors (Beck, 1972, 1976; Cormier & Hackney, 1993; Ellis, 1979; Ellis & Bernard, 1986; Meichenbaum, 1977) for certain ideas about decision making and reinforcement of positive action.

Each of the theoretical positions seems to have special merit in describing some aspect of the counseling process. Although the proponents of each system tend to believe that their particular focus comprises all of the most important elements of counseling, we believe an aggregate of the several systems to be more complete than any one alone. Of course, we do not suggest that a counselor randomly select techniques or strategies. We have provided a basis for selecting particular techniques according to the stage of counseling and the nature of the client's concerns, allowing a consistent progression through the counseling process to be planned. Chapter 12 provides synopses of selected theories and describes how each theory contributed to our generic model.

FUNDAMENTAL PRECEPTS OF EFFECTIVE HELPING

The search to identify those common elements that form the basis for effective helping continues. This section presents a set of fundamental precepts that form the basis for our understanding of the helping process.* These precepts have emerged from four sources: our study of the range of counseling theories, our experiences as helpers, our experiences as counselor educators, and our collaboration as authors interested in describing the helping process to others. The precepts are intended to orient you to some of the principles we believe to be most important in understanding the helping process. These principles will take on deeper meaning as you assimilate subsequent chapters and should be studied again when you have finished reading this book.

Precept One: Understanding Human Behavior

To be truly effective, the counselor must have a thorough understanding of human behavior in its social and cultural context and must be able to apply that understanding to each client's particular set of problems or circumstances.

Diagnosis and hypothesis generating are critical and inevitable parts of a counselor's work. The process of diagnosis has two interrelated functions: (1) to describe significant patterns of cognition, behavior, or affective experience; and (2) to provide causal explanations for these significant patterns. The process is one of developing tentative hypotheses, confirming their validity, and using them as the basis for making critical decisions concerning the focus, process, and directions of the counseling experience. The process of arriving at a diagnosis is a mutual one in which the client and the counselor work together to identify significant patterns and their roots in the client's experience.

---

*This set of precepts has been adapted from Chapter 1 of *Helping Clients with Special Concerns,* by S. Eisenberg and L. E. Patterson (1979). Boston: Houghton Mifflin. Reissued (1990). Prospect Heights, IL: Waveland Press.

Understanding human behavior means having a set of concepts and theories that help one account for and explain significant human reactions and relate them to experiences. These concepts and principles provide the core for the counselor's diagnostic and hypothesis-generating work. Counselors use theories and concepts about human behavior to understand their own behavior as well as the concerns, actions, perceptions, emotions, and motivations of their clients.

There are two dangers involved in the diagnostic and hypothesis-generating process. One is that the process often becomes a game of applying labels to clients, thus putting them into categories. Once categorized, a client is stereotyped; all of the general characteristics of those in the category are attributed to the client. As a result, the client's uniqueness as an individual may be lost. Worse, other important attributes of the client are missed because categorizing creates perceptual blinders for the counselor. A second danger is that helping professionals often make mistakes in their diagnoses, and these mistakes often result in ineffective and sometimes counterproductive helping efforts.

We agree that these dangers are real. But they are not inherent in the diagnostic process itself; rather, they are dangers arising from the misuse of the process. Counselors who comprehend the role that an understanding of human behavior and its social and cultural context plays in their work and who recognize the proper function of diagnosis will work very hard to avoid these dangers. It is part of their ethical responsibility.

Precept Two: Change in the Client

The ultimate purpose of the counseling experience is to help the client achieve some kind of change that he or she will regard as satisfying.

Virtually every significant theory of counseling states that creating growth-oriented change in the client is the ultimate intended outcome of the counseling experience. Some say that overt behavior change is the *sine qua non* of the experience. Others say that behavior change is just symptom change; real and lasting change comes when the client develops new perceptions of self, significant others, and life. Furthermore, some counselors take a remedial approach; they attempt to help the client change dysfunctional behavior into more functional patterns, such as by overcoming shyness, reducing debilitating anxiety, controlling counterproductive anger, or reducing interpersonal conflicts. Others believe that the goal of counseling is to help people make important life decisions; in this case, the counselor's role is to help the client use a rational thinking process to resolve confusion and conflict. Still other counselors view their work as stimulating favorable personal and interpersonal growth. As they see it, remediating dysfunctionality and assisting in decision making are just some of the means to the end of personal growth. For these counselors, helping people to become more complete or fully functioning is the crucial outcome of the counseling process.

Client change is often difficult to document. Behavior change, if it occurs, is probably the easiest to observe because it is the most tangible. However, clients may also change their views about certain behaviors that they previously regarded as undesirable, change the extent to which they experience stress related to an unwanted situation, or reduce their general level of emotional distress. In spite of the difficulties of assessing some kinds of change, a counselor who cannot describe the changes that a client has undergone has no basis for knowing when counseling has reached an effective conclusion.

Precept Three: The Quality of the Relationship

The quality of the helping relationship is significant in providing a climate for growth.

As each person meets another person, he or she decides how much of himself or herself to share with that individual. Before sharing deeply, we all assess the risk in terms of how we think the listener will react to our personal thoughts and feelings. If trust is not present, we remain closed and go away from the experience without benefit.

The critical elements of the helping relationship that promote openness are described often in the literature of the field: respect (rather than rejection), empathy (rather than shallow listening and advice giving), congruence or genuineness (rather than inconsistency), facilitative self-disclosure (rather than being closed), immediacy (rather than escapism to past or future), and concreteness (rather than abstract intellectualizing). Counselors must communicate respect for clients as persons with rights who are trying to live the best lives they can. Genuine caring is communicated when counselors try to understand the client's world as though it were their own and give the client verbal cues about that understanding. Effective counselors share their reactions to the client with the client, using this feedback as a way of helping the client toward deeper levels of self-understanding. Such counselor self-disclosure must always focus on the needs and goals of the client, not on the counselor's emotional needs.

Precept Four: A Sequential Process

Counseling is a process that occurs in a fairly predictable sequence and that is characterized by movement toward identifiable outcomes.

Counseling has a beginning, a middle, and an end. At the outset, the client and the counselor discuss the client's concerns. The counselor attempts to learn as nearly as possible what the client is experiencing and what has brought him or her to counseling. As the counselor listens to the client, he or she works to develop an appreciation of the client's phenomenal world: how the client perceives self, significant others, and surrounding life space. This stage of counseling involves more than identifying a problem or concern; it also includes developing a knowledge of how the client experiences that problem or concern in the world as he or she sees it. As the client discloses information about his or her concerns, those concerns come into sharper focus through the telling. Understanding the client's phenomenal world enables the counselor to offer constructive confrontations that can lead the client to greater self-awareness, an essential quality of the second stage of the counseling experience.

In the second stage, the client undergoes an in-depth exploration process in which the counselor tries to help the client understand self and environmental conditions that relate to the concerns that the client has chosen to work on. Intense listening, careful reflecting, and occasional confrontation characterize the counselor's contribution to this stage of the process. Out of the process of in-depth exploration comes a mutually agreeable diagnosis of the problem(s) and an understanding of the experiences that led to the current difficulties. Implicit in the process of diagnosis is the identification of goals for change. Once the client and counselor have a clear sense of what has gone wrong, they begin to turn their attention to alternative patterns of response. New insights, discoveries, and awarenesses about self characterize the client's experiences during this stage of the process. The experience is exciting for some, uncomfortable for others.

In the third and final stage of counseling, client and counselor join together in identifying specific goals for change and then review alternative plans to achieve these goals. Once action plans are formulated, the client begins implementing the changes, and client and counselor monitor the success of the action plans. Plans that are not helpful are eliminated, and new ones are substituted. Finally, when the goals are achieved to the client's satisfaction,.counselor and client bring the counseling process to a positive termination.

Precept Five: Self-Disclosure and Self-Confrontation

The counseling process consists primarily of self-disclosure and self-confrontation on the part of the client, facilitated by interaction with the counselor.

For counseling to take place, the client must disclose personal information to the counselor, who in turn tries to understand the client's world in a context of what he or she knows about how people respond to life situations. Although clients may reveal significant personal information through their nonverbal behavior, communication in counseling is primarily verbal. Clients reveal their thoughts and feelings to a perceptive counselor by what they say, the affect with which they say it, and what they choose to obscure in their verbalizations. The more fully self-disclosure takes place, the more effectively the counselor can help the client discover new ways of coping. For some clients self-disclosure comes easily, and for others it is more difficult. Counselors need to be aware of the potential difficulty of self-disclosure, especially for clients of a different culture or for whom the family of origin did not foster such activity. In such circumstances, the counselor must be particularly attentive to establishing trust and must have a flexible repertoire of skills for facilitating self-disclosure.

Self-confrontation by the client is a process of looking at self with an expanded perspective that allows for the development of new perceptions about self. The counselor assists the client to broaden his or her perspective on self by providing honest feedback. At the simplest level, that feedback may simply be restatements of the client's own words—restatements that cause the client to reconsider a thought just expressed. As the counselor becomes more confident of his or her understanding of the client, he or she may choose to move to a more comprehensive form of feedback that helps the client see himself or herself in a situation from an alternative viewpoint. Since such feedback comes from the counselor's frame of reference, it frequently will be a view that the client has not previously considered. It is important for the counselor to be as free of vested interest as possible in using confrontation as a counseling tool.

Whether the counselor's feedback is at the low-risk restatement level or the higher-risk confrontation level, the client must confront self with new ways of seeing and understanding self in life situations. Through this process, a new understanding of personal needs, desires, perceptions, assumptions, and cognitions emerges, and new coping skills are discovered and refined.

Precept Six: An Intense Working Experience

Counseling is an intense working experience for the participants.

For the counselor, the related activities of attentive listening, information absorption, message clarification, hypothesis generation, and treatment planning require sustained energy. Beyond these largely intellectual activities is the emotional experience of caring for another person enough to be affected by that person's emotions without becoming lost in those emotions and therefore diminished as the facilitator.

For the client, the hard work comes in the effort to understand what is difficult to understand; in the endurance of confusion, conflict, and uncertainty; and in the commitment to disclose to self that which is painful to think about. This effort, endurance, and commitment require a level of concentration that may never have been experienced before. All clients undergo the added stress of revealing personally felt inadequacies to another. The work of producing growth through counseling is always demanding for the client and often painful, though at the same time fulfilling and rewarding.

Counseling is not the same thing as conversation. In conversation, two or more people exchange information and ideas. The experience is usually casual and relaxed. People leave a conversation and move easily to other things. Counseling, on the other hand, is characterized by a much higher level of intensity. Ideas are developed more slowly, encountered at a deeper personal level, and considered more carefully. People often leave a counseling session mentally and emotionally depleted, yet still thinking about what was discussed. At other times, people are energized by the new insights they have gained—in spite of the painful aspects of the new awareness.

Precept Seven: Ethical Conduct

Offering to provide professional people-helping service obligates the helper (counselor, social worker, psychologist, or other) to function in an ethical manner. Codes of ethics published by the relevant professional associations serve to set some needed parameters.

Ethical practice may be defined as providing, with care and conscientious effort, a helping service for which one has been appropriately trained. Unethical practice occurs when counselors practice outside the limits of their competence, fail to place their clients' interests ahead of their own needs, or fail to respond sensitively to their clients' life experiences and rights. Because counselors present themselves to the public as persons with special skills for helping people in need, they have a greater burden not to do harm than do other citizens who do not purport to be expert helpers. Counselors need to be aware of the great responsibility they take on when they provide services to clients. Counseling that is incompetent or insensitive or that serves the interests of the counselor not only causes harm to the clients who receive it but also damages the reputation of the counselor's employer and the profession as a whole.

Codes of ethics (see Appendixes A and B) serve to answer many questions about ethical dilemmas that present themselves in counseling practice. Understanding the ethical principles underlying the codes and broader ethical theories is necessary for resolving some complicated ethical dilemmas. In the long run, the responsibility for ethical action always rests with the judgment of the individual practitioner.

CHARACTERISTICS OF EFFECTIVE HELPERS

The purpose of the following exercise is to help you identify the important qualities and characteristics of effective helpers. Although it can be done in private, it may have more meaning if shared with others participating in the same learning experience.

Exercise

Step One

During your life, you have had experiences in which you were a receiver of help; you have also had experiences in which you were the provider of help. Whether or not these experiences were labeled as counseling, at important points in your life you have both given and received help. For this exercise, recall two occasions (involving different people) when you sought help from another. In either instance, the helper might be a friend, spouse, supervisor, parent, teacher, or counselor. The first occasion should be one in which the help you received was valuable to you; that is, the helper was effective in offering help to you. The second occasion should be one that did not result in success; you did not receive the help you needed.

Step Two

Think of the first experience and write down your responses to the following questions:

- What kind of help were you seeking? (What were you concerned about? What did you hope to accomplish as a result of your discussion?)
- How did the helping person treat you? What were his or her basic attitudes toward you and your concerns?
- What response patterns did you notice in the person who offered help to you?
- What were the most memorable characteristics of the person who offered help to you?

Step Three

Respond to the above questions again with the second (less successful) helping experience in mind. Do not proceed until you have completed this step.

Step Four

Compare and contrast the two helping experiences and the helpers who were involved. What factors were present during the first helping experience that were missing during the second? How did the helpers differ from each other? Write down these comparisons. Again, please do not proceed until you have completed this step.

Step Five

Once you have compiled your comparisons, share your findings with other students. Then see whether you can verify from your experiences the observations that follow.

Through the years, the characteristics of effective helpers have been among the most popular dissertation subjects, and there is also an extensive research literature on this subject by established scholars. Most studies attempt to relate particular characteristics, such as dogmatism, to counselor effectiveness. Because counseling is so complex, each study contributes but a small part to the total picture of what makes an effective counselor. Counseling theorists and practitioners have also added their clinical observations about the characteristics of effective helpers. This section summarizes several of the most important qualities possessed by effective helpers.

Effective helpers are skillful at reaching out. Through their demeanor and underlying views about others, effective helpers are able to encourage others to communicate openly and honestly with them. They avoid responding in ways that create defensiveness and block communication. Effective helpers do this by participating in active and involved listening. They are able to concentrate fully on what is being communicated to them, not only to understand the content of what is being said but also to appreciate the significance of that verbalization to the client's present and future well-being. Effective helpers listen actively for feelings, beliefs, and perspectives and assumptions about self, significant others, and life circumstances. They are able to control their own feelings of anxiety while hearing of another person's concerns and anxieties.

Effective helpers inspire feelings of trust, credibility, and confidence from people they help. In the presence of effective helpers, clients quickly sense that it is safe to risk sharing their concerns and feelings openly. They will not be ridiculed, embarrassed, made to feel ashamed, or criticized for the thoughts, feelings, and perceptions they share. Nothing "bad" will happen as a consequence of sharing, and there is a very real chance that something gainful will come of it. Effective helpers are also credible. What they say is perceived as believable and honest. Furthermore, effective helpers do not have hidden agendas or ulterior motives. They are viewed as honest, straightforward, and nonmanipulative, further supporting their clients' belief that they can be trusted.

Effective helpers are able to reach in as well as reach out. Effective helpers do a lot of thinking about their actions, feelings, value commitments, and motivations. They show a commitment to nondefensive, continuous self-understanding and self-examination. They are aware of the feelings they experience and the sources of those feelings. They are able to manage anxiety by being aware of it and its sources rather than blotting it from awareness. Effective helpers are able to respond with depth to the question "Who am I?" They can help others think openly and nondefensively about themselves and their concerns because they are not afraid to participate in such experiences.

Effective helpers communicate caring and respect for the persons they are trying to help. By their demeanor, effective helpers communicate to their clients the following unspoken statement: "It matters to me that you will be able to work out the concerns and problems you are facing. What happens to you in the future also matters to me. If things work out well for you and you achieve success, I shall be happy about it. If you encounter frustration and failure, I shall be saddened." The opposite of caring is not anger, but indifference. Effective helpers are not indifferent to the present and future of the people they try to help. On the contrary, effective helpers agree to offer time and energy to others because the future well-being of the people to whom they are reaching out matters to them.

To respect another person means to hold that person in regard and esteem. It means to have a favorable view of that person—to acknowledge his or her talents and not think less of him or her because of any limitations. Applied to effective helping, respect means believing that the client is capable of learning, of overcoming obstacles to growth, and of maturing into a more responsible, self-reliant individual. With this perspective, effective helpers communicate regard for others by offering their time and energy and by listening in an active, attentive fashion that shows involvement. They also show respect by not treating the individual as though he or she were stupid or ridiculous, incapable of sensible thinking and reasoning. Effective helpers respect the role of social factors and cultural heritage in each client's experiences and actively work to express their caring and respect in ways that are consistent with the client's cultural and social background. Effective helpers also like and respect themselves, but they are neither arrogant nor conceited and thus do not show arrogance toward the people they are helping.

Effective helpers like and respect themselves and do not use the people they are trying to help to satisfy their own needs. Every human being wants to be accepted, respected, and recognized by significant others and to be acknowledged for his or her special talents and achievements. However, some people are especially dependent on others for recognition and acknowledgment. They purposely respond to "you're okay and likable" feedback messages from others, and their responses tend to satisfy their own needs. People who do this excessively eventually "turn off" other people and make them afraid. This interpersonal pattern blocks honest communication and leads instead to game playing. Truly effective helpers feel secure about themselves and like themselves and thus are not dependent on the people they are trying to help for respect, recognition, and acknowledgment.

Counselors who are under great stress in their personal lives are at risk of focusing on their own needs rather than their clients' during counseling sessions. Thus they ought to monitor carefully their effectiveness with clients. If personal stress is compromising a counselor's effectiveness, then the counselor needs to seek help and arrange for alternative care for his or her clients until the stress is reduced. A counselor who is needy and unable to focus on the client's best interests is commonly referred to as an "impaired" counselor or a "wounded healer." Under such circumstances, counselors can do more harm than good. They have either temporarily succumbed to overwhelming stress or have long-standing emotional difficulties that prevent them from relating positively to others. The most common reason for professional impairment is alcohol or drug dependence.

Even counselors who have worked through their own emotional difficulties or who are not overwhelmed by stress in their personal lives are at some risk for impairment. They can develop burnout, feelings of emotional depletion and alienation from clients and a sense of futility in their work. Burnout happens when counselors let their work become the only focus in their lives and when they work under conditions that make the job itself more stressful. This combination of circumstances puts counselors at risk for losing perspective on their effectiveness and their clients. To avoid burnout, counselors must take time away from work to care for themselves, nourish their own personal support systems, and get a clearer perspective on their accomplishments as counselors. Attending to day-to-day working conditions and acting to make them as stress-free as possible also reduce the risks of burnout.

Effective helpers have expertise in some area that will be of special value to the person being helped. Employment counselors have special knowledge about the career development process, the skills needed for decision making, and jobs available in their local community. Pregnancy counselors have special knowledge about available clinics, the functioning of the human body, laws and philosophical and religious beliefs pertaining to abortion, and human sexuality. Counselors who work with children have special knowledge of child development, special counseling tools that are effective with children, and special skill in family counseling. When people need help, they turn to people who they believe have knowledge about and expertise in the area of concern to them. When faced with a personal problem, they will turn to whomever they can identify as having an especially sound knowledge of human behavior.

Effective helpers attempt to understand the behavior of the people they try to help without imposing value judgments. There appears to be a tendency for people to make value judgments about the behavior of others—to judge the behavior of others by their own standards. Though appropriate when casting a vote, this value-judging tendency seriously interferes with the process of effective helping. Effective helpers work hard to control the tendency to judge the values of their clients. Instead, they accept a given behavior pattern as the client's way of coping with some life situation, and they try to understand how the pattern developed. The effective helper will develop opinions about whether or not the behavior pattern is serving the client's goals or best interests but will refrain from classifying the client's values as "good" or "bad."

Effective helpers develop an in-depth understanding of human behavior. They understand that behavior does not simply occur. Their approach is that all behavior is purposeful and goal-directed, that there are reasons and explanations for human behavior, and that truly helping another means understanding the reasons for that person's behavior rather than judging them.

Effective helpers are able to reason systematically and to think in terms of systems. A *system* is an organized entity in which each of the components relates to each other and to the system as a whole. Examples of systems include the human body, the organizational setting in which a person works, and the family unit. In "high-entropy" systems, components work cooperatively with each other and contribute favorably to the goals of the total system. In "low-entropy" systems, components do not work cooperatively, and sometimes work against each other. Effective counselors are aware of the different social systems of which a client is a part, how he or she is affected by those systems, and how he or she, in turn, influences those systems. In other words, effective helpers are aware of the forces and factors in a client's life space and the mutual interaction between the client's behavior and these environmental factors. Effective helpers realize that a client's concerns and problems are influenced by many complex factors that must be identified and understood as an inherent part of the helping effort.

Effective counselors are contemporary and take a world view of human events. Counselors are aware of important present-day events in all the systems affecting their lives and the lives of their clients. They are aware of the significance and possible future implications of these events. Being contemporary means that the counselor has a thorough understanding of current social concerns and an awareness of how these events affect the views of clients—especially about the future. The converse of being contemporary is being encapsulated—unaware of what is happening in the systems and environments that make up one's life space (Wrenn, 1973). Among the important contemporary issues to which a counselor must attend is how bias and discrimination against some groups in society affect their members' personal well-being and progress toward self-actualization.

Effective helpers are able to identify behavior patterns that are self-defeating and to help clients change self-defeating behaviors to more personally rewarding behavior patterns. People frequently do things that are counterproductive and goal-disruptive rather than goal-enhancing. Some communicate ridicule and hostility when they want respect and friendship. Others run away from frightening situations rather than confronting the aspects of a situation that cause anxiety. Still others do things to betray trust and cannot understand why others do not trust them. And others are afraid to respond assertively when people make unreasonable demands on them. Counselors assist by helping clients recognize that they are repeating behaviors that lead to unwanted outcomes. Then alternative plans are developed and rehearsed so that clients can begin to behave differently.

Effective helpers are able to distinguish between healthy and unhealthy behavior patterns and to aid others in working toward the development of healthy, personally rewarding patterns. Effective helpers have a model, or image, of the qualities and behavior patterns of a healthy and effective, or fully functioning, individual. Included in this model is a pervasive awareness of effective and ineffective ways of coping with the stressful situations of life. Effective helpers are skillful at helping clients look at themselves and respond nondefensively to the question "Who am I?" It is easy to describe aspects of self that are likable and admirable. It is difficult and painful to look at those aspects that are not. Yet self-improvement and growth require an honest, open awareness of those aspects of self that one would like to change. Effective counselors are able to help clients to look at themselves—at both their likable and less admirable aspects—without debilitating fear, identify personal changes that would promote growth and improvement, and develop approaches to bring about those improvements.

Exercise Continued

This section described eleven characteristics of effective helpers. How did they compare to the characteristics you identified in the earlier part of this exercise? Did you identify some characteristics that were not discussed in the material you just read? Your observations will contribute to your personal image of effective helping. Were any characteristics discussed that you did not identify? Reading about the ones you did not see should cause you to think about how those qualities might have fit your personal helping situation.

Some of the characteristics discussed may have seemed obvious to you, and others may have seemed controversial. You probably observed, for example, that terms such as *empathy, genuineness, positive regard, self-disclosure,* and *concreteness* were not used. However, if you observe closely, you will find these concepts presented in other words, such as *reaching out, credible, reaching inside of self,* and *reasoning systematically.* As you consider the characteristics of effective helpers you have known, you will find that everyday language may be used to describe the effective helper.

As the final step of this exercise, reflect on your personal qualities as they relate to being an effective helper. Do your qualities resemble those of effective helpers you have known and those described in this section? Do you see special strengths? Do you see limitations that may be remedied through study and practice? Do you see limitations that might cause you to question your choice to become a professional helper? If you experience confusion or stress about your potential to be effective in this work, discuss your concern with your professor or a qualified professional helper.

THE JOY OF HELPING

Throughout this book, we repeat over and over that counseling must always serve the purposes of the client, that the focus of the talk must be on the client's issues, and that the counselor should make every effort to be intellectually and affectively available to the client throughout the process. We would be remiss if we did not also address the great satisfaction that comes from helping others.

There is the fascination of learning firsthand how others experience the world, indeed, how two people can participate in the same event and experience it differently. There is the wonder of seeing the human spirit in action as clients overcome adversity and grow toward new meanings in their lives. There is the great joy that comes from being the catalyst in helping clients escape from the miseries of emotional constriction and distress, indecision, and copelessness.

It is a privilege to be allowed to share the private worlds of others, not only because it is essential to the helping process but also because it enriches one's life as a counselor. We encourage you to reflect on the rewards that come to the helper in the counseling process and to find a personal sense of worth in the work that is accomplished. Although counseling is for the client, the counselor gains, too.

In Chapters 2 through 7, we present a detailed description of our three-stage generic model of counseling. Technical terminology is introduced so that the reader can relate this work to the broader literature in the field. We describe counseling skills involved in building trust; structuring and leading client self-exploration; diagnosing problems; and goal setting, action planning, and termination. Chapters 8 through 11 explore the ways in which the counseling process needs to be refined to accommodate special client populations: reluctant clients, clients in crisis, children, women, and persons of different cultures. In Chapters 12 and 13, we review how theoretical orientation affects the counseling process and discuss the important ethical issues facing counselors.

SUMMARY

This chapter presented the theoretical background of our thoughts about effective helping, some fundamental precepts that describe the essential elements of the counseling process, and a description of the qualities of the effective helper. Inevitably, these qualities derive from and feed into the fundamental precepts of effective helping, because the counseling process cannot be understood apart from the person of the helper. Finally, those who are thinking of becoming professional helpers are urged to take an introspective look at themselves in order to identify the presence or absence of qualities that have been identified with effective helping. It has been our experience that people who want to tell others how to live effectively are rarely good counselors; people who want to help others gain control over their own lives usually do well as counselors and experience personal satisfaction in their work.

REFERENCES

Alexander, F. M. (1963). *Fundamentals of psychoanalysis.* New York: Norton.

Beck, A. (1972). *Depression: Causes and treatment.* Philadelphia: University of Pennsylvania Press.

Beck, A. (1976). *Cognitive therapy and emotional disorders.* New York: International Universities Press.

Carkhuff, R. R. (1969a). *Helping and human relations* (Vol. 1). New York: Holt, Rinehart and Winston.

Carkhuff, R. R. (1969b). *Helping and human relations* (Vol. 2). New York: Holt, Rinehart and Winston.

Cormier, L. S., & Hackney, H. (1993). *The professional counselor* (2nd ed.). Boston: Allyn and Bacon.

Egan, G. (1990). *The skilled helper: A systematic approach to effective helping* (4th ed.). Pacific Grove, CA: Brooks/Cole.

Eisenberg, S., & Patterson, L. E. (1979). *Helping clients with special concerns.* Boston: Houghton Mifflin. Reissued (1990). Prospect Heights, IL: Waveland Press.

Ellis, A. (1979). Rational-emotive therapy. In R. J. Corsini (Ed.), *Current psychotherapies* (2nd ed.). Itasca, IL: Peacock.

Ellis, A., & Bernard, M. E. (1986). What is rational-emotive therapy (RET)? In A. Ellis & R. Grieger (Eds.), *Handbook of rational-emotive therapy: Vol. 2* (pp. 3–30). New York: Springer.

Meichenbaum, D. (1977). *Cognitive behavior modification: An interactive approach.* New York: Plenum.

Perls, F. (1969). *Gestalt therapy verbatim.* Moab, UT: Real People Press.

Polster, E., & Polster, M. (1973). *Gestalt therapy integrated.* New York: Brunner/Mazel.

Rogers, C. R. (1942). *Counseling and psychotherapy.* Boston: Houghton Mifflin.

Rogers, C. R. (1951). *Client-centered therapy.* Boston: Houghton Mifflin.
Rogers, C. R. (1961). *On becoming a person.* Boston: Houghton Mifflin.
Rogers, C. R. (1980). *A way of being.* Boston: Houghton Mifflin.
Rogers, C. R. (1986). Carl Rogers on the development of the person-centered approach. *Person-Centered Review, 1,* 257–259.
Strachey, J. (Ed.). (1964). *The standard edition of the complete psychological works of Sigmund Freud.* London: Hogarth.
Williamson, E. G. (1939). *How to counsel students.* New York: McGraw-Hill.
Williamson, E. G. (1950). *Counseling adolescents.* New York: McGraw-Hill.
Williamson, E. G. (1965). *Vocational counseling.* New York: McGraw-Hill.
Wrenn, C. G. (1973). *World of the contemporary counselor.* Boston: Houghton Mifflin.

Chapter 2

Understanding Counseling as a Process

Counseling is a term that connotes many different activities. The word is commonly applied to the activity of attorneys, insurance counselors, and even cosmetologists. Implicit in each use is a definition of counseling that means advice giving. The early practice of counseling in the schools was indeed advice giving. In the beginning of its history, school counseling focused heavily on telling the young and inexperienced what they ought to do. Subsequent to Carl Rogers's publication of *Counseling and Psychotherapy* (1942), counseling as a helping service began to change and to ally itself with psychology and social work, while retaining some of its historical ties to education. Rogers's germinal work suggested that perhaps one person's solutions may not fit another person's capacities, values, or goals and that being an effective helper requires a thorough knowledge of the client. Since different people might develop different responses to the same situation and be equally satisfied with the results, the role of advising and teaching in counseling was drastically reduced. The focus of counseling turned to helping people clarify their own goals and build plans of action accordingly. Such an approach emphasizes the unique potential of each individual and defines the role of the counselor as a facilitator of personal growth. Rogers's definition now forms the foundation for current definitions of professional counseling.

The term *psychotherapy* is closely related to *counseling* as we use the word in this book. Psychotherapy has its roots in the medical and psychological tradition that sees the suffering that brings people to professional helpers as mental illness, or sickness. In time, psychotherapy has come to denote a much wider range of activities than treatment of the mentally ill, although some focus on mental illness is still retained. As described in the current literature, the outline of the process of psychotherapy closely parallels the model in this book. We have retained the word *counseling* instead of using *psychotherapy* because we believe that the former omits the connotation of an exclusive focus on mental illness and because it more easily includes activities in educational as well as mental health settings.

Since professional counseling as a helping activity is much more complex than the conventional use of the word, we need to provide a comprehensive definition of counseling. The definition that follows addresses both the outcome goals and the process goals of counseling. A discussion of the goals follows the definition and serves as an orientation to Chapters 3 through 7, which describe the stages of counseling in detail.

DEFINITION OF COUNSELING

Counseling is an interactive process characterized by a unique relationship between counselor and client that leads to change in the client in one or more of the following areas:

1. Behavior
2. Beliefs (ways of construing reality, including self) or emotional concerns relating to these perceptions
3. Ability to cope with life situations so as to maximize opportunities and minimize adverse environmental conditions
4. Decision-making knowledge and skill
5. Level of emotional distress

In all cases, counseling should result in free and responsible behavior on the part of the client, accompanied by an ability to understand and manage his or her negative emotions.

OUTCOME GOALS OF COUNSELING

Change Must Occur

The first element of the above definition that alludes to goals is that counseling "leads to change in the client." This is true of either individual counseling or group counseling and whether the expressed intent of the counseling is developmental (oriented to personal growth) or remedial (oriented to the resolution of problems). The change that occurs may be overt and dramatic, or it may be imperceptible to anyone but the client. Awareness that counseling is supposed to lead to change diminishes the temptation to think of counseling as "just a nice conversation" and sets the tone for the hard and sometimes painful work that effective counseling requires. Successful counseling can result in gratifying, even exhilarating, changes for the client, but these positive results don't happen by magic or without real commitment. Comprehension of this fact is fundamental to change. There are many ways in which clients may change as a result of counseling; the following discussion of the scope of possible change will reveal that change of some kind is a reasonable outcome goal.

Categories of Possible Change

Change in counseling can take several forms: behavior change, modification of beliefs or values, improvement in decision-making or coping skills, or reduction of the level of emotional distress. This section examines each category of change, beginning with behavior change.

Behavior change is probably the easiest type of change to recognize because it is overt and observable. A behavior change might be the solution of a problem, as in the case of a child who learns to get what he wants from others through verbal requests and negotiation rather than through fighting. A behavior change might also enhance one's potential for personal growth, as in the case of a middle-aged person who returns to school or embarks on a new career. Many counselors believe that changes in thoughts and attitudes must precede changes in behavior, and they work to understand those changes. Counselors of the traditional behaviorist school maintain that a counselor can never really know a client's inner thoughts and attitudes and that only observable behavior changes serve to indicate the success of counseling. Modern behaviorists tend to consider changes in thinking as a mediating factor in behavior change.

Though not directly observable, a change in beliefs (also called personal constructs) may occur in counseling and can be assessed from the verbal output of the client. A common goal of counseling is that the client will improve his or her self-concept and come to think of himself or herself as a more competent, lovable, or worthy person. Kelley (1955) describes personal constructs as an individual's particular view of reality (Patterson, 1986). Kelley states that people's behaviors are based on what they believe to be true. Therefore, people who think they are incapable and feel embarrassed about performing in front of others will act on those personal constructs by avoiding anything challenging. Changes in beliefs often lead to behavior change, but they can also lead to changed views that make present behavior more satisfying. Ellis (1989) has developed a system for understanding how one's personal thoughts can lead to dissatisfaction with the state of one's life. He explains that people acquire irrational thoughts that lead to expectations that can never be fulfilled, for example, "I must be perfectly competent in all that I do, or I am not a good person." Through counseling, the client may learn to give up such thinking and instead appreciate what he or she does well while working toward competence in other areas. Beck (Beck, 1976; Beck, Rush, & Emery, 1979) presents a similar model that describes the impact of negative belief systems on feelings and behaviors. Meichenbaum (1977) refers to this process as the internal dialogue and defines one goal of counseling as modifying the content of the dysfunctional internal dialogue.

In the case study that follows, the client eventually changes both his thinking and his behavior. Observe how his changes in thinking free him to behave more effectively.

Case: Thad

Thad, a young man of 22, sought a counselor at the university counseling service because he had trouble concentrating on his studies and felt constant tension. Two months earlier, he had moved into his own apartment, leaving his parents' home for the first time. He felt guilty about "deserting his parents when they needed him." His father had a serious degenerative health problem and required some special care, though he was not an invalid. His mother was a vigorous and healthy woman who was able and eager to help make her husband's life comfortable.

As counseling proceeded, Thad came to understand that his parents needed his affection and involvement but not his physical help and daily presence. His daily visits were actually interrupting other things they wanted to do. In other words, his belief that he was needed in his parents' home each day or he could not call himself a good son was amended by his realization that his physical presence was not the critical factor in their perception of him as a good son. Thad changed his personal perception of his role in the family, and his new perception allowed him to live away from home without guilt. He began to make special occasions of his visits to his family, and he and his parents began enjoying each other's company once more. Thad's feelings of tension receded and he was able to refocus on schoolwork and other elements of his personal life.

Questions for Further Thought

1. What other irrational thoughts may have led to Thad's feelings of guilt about his treatment of his parents?
2. What rational thoughts do you believe replaced them as counseling proceeded?
3. How did Thad's changed thinking affect his behavior?
4. How do you think this new perception of self and family might affect his other relationships?

Counseling may also enhance an individual's ability to cope with life situations. Certain environmental conditions are adverse and difficult to change, but learning how to manage one's life in the face of adversity creates room for accomplishment and enjoyment in spite of such conditions. For example, some people with terminal illness refer to the period after they got sick as one of the best of their lives because of the closeness to and honesty with loved ones that their impending death brought. Clearly, they are not glad they got sick; rather, they mean that they are able to appreciate the precious gains the illness provided, in spite of its devastating consequences.

Coping ability depends upon the individual's skill in identifying the questions to be resolved, the alternatives that are available, and the likely results of different actions. Sometimes coping means learning to live with what one cannot change. For example, a person who has suffered a permanently disabling injury in an automobile accident may use counseling to learn to cope with his or her new situation.

Learning to cope with an alcoholic parent requires that certain questions be resolved: What can I change, and what must I accept? Who can help me? Am I responsible for my parent's behavior and well-being? What do I want to do? What are the likely results of available actions? A counselor can help by encouraging the client to define dimensions of the family system as a basis for predicting what may work and what risks are worth taking. Depending on the analysis, the conclusion might range from staying away from the parent when he or she is drunk to arranging for the parent to seek treatment for his or her problem. Changes in ability to cope usually include both behavior change and change in personal constructs.

Counseling may also contribute to a client's ability to make important life decisions. The counselor teaches the client self-assessment procedures and how to use information to arrive at personally satisfying answers. Career decision making is still a major focus of school and college counselors. Counselors prepared in contemporary career development methods focus heavily on helping clients identify relevant sources of information. They generally refrain from giving advice and see career decision making as a lifelong process rather than a single decision made during young adulthood.

An additional function of counseling is the relief of emotional distress (Brammer, Shostrum, & Abrego, 1989). Many clients enter counseling because they feel bad and need a place where they can safely vent those feelings and feel sure that they will be accepted and understood. Their level of emotional distress may be interfering with their daily activities, and they need relief from their psychic pain. Many times, the relief of emotional distress is only one piece of the necessary change, and attention to irrational thoughts, inadequate coping skills, or dysfunctional behaviors that perpetuate the emotional distress is critical to lasting change. At other times, individuals have been overwhelmed by a loss or tragedy in their lives and benefit largely from the emotional release. A person whose house has been destroyed in a hurricane may appropriately use counseling to vent the fear experienced during the storm and the sad and angry feelings about the losses due to it. Once the trauma has been processed, pre-existing coping skills may be sufficient for moving on in life.

Change that occurs in counseling can influence feelings, values, attitudes, thoughts, and actions. Among the broad variety of potential changes, some will be obvious and others very subtle. Since the scope of possible change covers essentially all dimensions of human experience, it can correctly be stated that if change in at least one dimension does not occur, counseling has not succeeded. The result of counseling may be inner peace—with little outward sign of change. At other times, behaviors will change markedly, and the client may then need help in understanding the reactions of others to the changed behavior.

Free and Responsible Behavior

Freedom is the power to determine one's own actions, to make one's own choices and decisions. Throughout their history, humans have migrated from one location to another in search of a social order that would allow freedom, and this nation was founded by people who crossed an uncharted ocean in search of freedom. But freedom is fragile, and some of it must be sacrificed as the price for living in any kind of social system. Freedom is limited by the responsibility to consider the freedoms of others as one determines one's actions; it is not license to do exactly as one pleases.

One role of counselors is to help clients assess the true margins of their freedom by focusing their thoughts on the consequences of their actions and decisions. Clients who feel that freedom is license must be helped to see that family, friends, teachers, employers, or the society at large will exact a price for behaviors that are perceived as threatening to the client's self-interest or the interests of others. Other clients may be all too willing to give away their freedoms to others (especially parents or spouses), abdicating the right to make decisions that will have major effects on the rest of their lives for the short-term gain of pleasing others. In either case, the counselor works with the client to come to a more balanced perception of personal freedom.

Counselors who deal with children and adolescents face conflicting guidelines about how far they may go in supporting free choice. The issue becomes one of children's rights versus parents' rights, in a context where both children and parents are capable of making very faulty judgments. Conventional belief associates wisdom with maturity, and so counselors may feel little support for respecting children's rights. The younger the child, the smaller the measure of autonomy he or she can reasonably be given. A counselor can help adolescent clients protect their freedoms by helping them consider alternatives carefully so that they can make well-examined decisions about their behaviors. Unfortunately, no absolute rule exists to help the counselor know when a client is sufficiently wise and mature to assume responsibility for self, and careful professional judgments must be made in each case. Counselors must be sure to respect their clients' freedoms and to help them show responsible respect for the freedoms of others.

Understanding and Managing Anxiety and Loss

It is a common misunderstanding that counseling eliminates anxiety. Beginning counselors are tempted to set the elimination of anxiety as one of their missions, and clients will reinforce them in this goal. The counselor need only look within self and to friends and family to realize that anxiety is present even in people who are leading satisfactory lives. It is definitely a goal of counseling to help people understand their anxieties and to reduce debilitating anxiety.

It is a useful convention to classify anxiety into reality anxiety, moral anxiety, and neurotic anxiety. Reality anxiety is the fear of real danger and is an essential safeguard to life. Moral anxiety is the fear of doing something that is morally wrong. Whether this anxiety is based on the fear of violating a code (such as the Ten Commandments or the Koran) or on the fear of violating someone else's rights, it serves as a control on people's social behavior and permits them to live with a certain amount of security. Neurotic anxiety is the fear of imagined dangers or humiliations and serves no useful purpose because it does not relate to reality.

Reality anxiety is rarely the subject of counseling, and a counselor will seldom attempt to change the level of reality anxiety exhibited by a client. A person ought to have sufficient anxiety about crossing a street to prompt checking for traffic. If one is a structural steel welder, it is a matter of self-preservation to observe safety precautions when building a skyscraper. On occasion, a counselor may attempt to teach a client that some action is dangerous if the client seems uninformed, as in the case of a child's playing with matches or an adolescent's failing to use a crash helmet when riding a motorcycle. Seeing excessive danger in situations of ordinary living is by definition neurotic anxiety, which is discussed below.

Moral anxiety, as previously stated, is an essential control in any group living structure. It serves as a deterrent to persons who are tempted to take another's property, privacy, or life. Persons who experience no moral anxiety are sociopathic. On the other hand, some people develop moral control that is so rigid that they cannot enjoy life. They worry for days if they think they have offended someone or if they have behaved in a way that violates their ideal of a moral person. As with reality anxiety, when moral anxiety becomes excessive, it may be classified as neurotic. In that case, it is debilitating and out of proportion with the need for real concern.

Neurotic anxiety is the main concern with many clients, since it creates anguish and severely inhibits effective functioning. An example of neurotic anxiety is fear of meeting new groups of people, when in fact one knows that one is quite successful in establishing new contacts; another is fear of riding an elevator, when one knows that elevators are usually safe. There is, of course, a judgment involved in deciding that such anxieties are neurotic, since some people are not successful in establishing new relationships and some elevators are out of repair and perhaps dangerous.

The goal of counseling is to leave clients with situation-appropriate levels of anxiety. Clients will then fear dangerous situations and will care appropriately about behaving morally. They will have eliminated most needless anxiety and will understand that the anxiety they do have is a motivator to deal with life's problems.

A second underlying issue that motivates many people to seek counseling is loss. Dealing with the pain of losing valued relationships, necessary jobs, health that has been taken for granted, or dreams and expectations that are unfulfilled is at the core of many presenting problems. Professional counseling exists to help people sort out the meaning of their lives in the face of such losses. Again, our approach is that counselors cannot impose their meanings or values on clients; rather, clients need to use counseling as a means to sorting out for themselves issues of meaning in spite of loss. Existential theories of counseling focus largely on apprehending meaning and purpose in spite of pain (e.g., Yalom, 1980).

PROCESS GOALS IN COUNSELING

Our definition said that counseling is an interactive process characterized by a unique relationship between counselor and client. The remainder of this chapter is devoted to an overview of the counseling process and an introduction to the nature of that special relationship. To understand counseling as a process, one must distinguish between outcome goals and process goals. Outcome goals (described in the previous section) are the intended results of counseling. Generally, they are described in terms of what the client desires to achieve as a result of his or her interactions with the counselor. In contrast, process goals are those events the counselor considers helpful or instrumental in bringing about outcome goals. Outcome goals are described in terms of change in the client that will be manifest after the counseling and outside the counselor's office. Process goals are plans for events that take place during the counseling sessions and in the counselor's office. They are events that the counselor considers helpful and instrumental to achieving outcome goals.

Process goals are sometimes described in terms of the counselor's actions and at other times in terms of effects to be experienced by the client. For example, a counselor may think, "If I am to help this client, I must actively listen to what he is saying and understand the significance of his concerns for his present and future well-being. I must understand how the attitudes he is describing influence the way he behaves toward significant others. I must understand the surrounding circumstances (including cultural background) that relate to his concerns, and I must understand the reinforcing events that support his behavior." All of these statements are process goals that relate to the counselor's behavior.

When reviewing a tape recording of the first session with a client, a counselor may think, "If I am to help this client, I believe he must feel a greater trust for me than he now appears to be experiencing. The client seems to be talking a good deal about issues and events that do not relate to his primary concerns. If our sessions are to be worthwhile, I think he must focus more intently on these concerns. If the client is afraid to think about or talk about them, it will be important to help him gain control over these anxieties. How can I help the client feel more trust and less anxiety?" Experiencing trust, focusing more fully on primary concerns, and controlling fear are all process goals described in terms of effects the client should experience. The last sentence in the counselor's thinking raises the crucial questions "What can I do?" and "How can I behave so as to facilitate these important process goals?"

Another kind of process goal relates to the way the counselor can act as a model for new ways of behaving. By modeling appropriate responses to frustration, disappointment, or negative feelings, the counselor indirectly teaches the client alternatives to accustomed ways of responding.

Some process goals appear essential for all counseling relationships and in fact define the steps in the counseling process. Other process goals are specific to particular clients. In the preceding example, concern about trust illustrates a common process goal in all counseling, whereas the client's avoidance of primary concerns is particular to him. Even though some other clients may show the same tendency, such behavior is not expected in all clients. Elements of the counseling process that are common to essentially all counseling interactions are introduced in the following discussion of the stages of counseling.

STAGES OF THE COUNSELING PROCESS

The word *process* helps to communicate much about the essence of counseling. A process is an identifiable sequence of events taking place over time. Usually there is the implication of progressive stages in the process. For example, there are identifiable stages in the healing process for a serious physical wound, such as a broken leg. Similarly, there are identifiable stages in the process of human development from birth to death. Although the stages in this process are common to all human beings, what happens within each of these stages is unique for each individual.

Counseling also has a predictable set of stages that occur in any complete sequence. Initially, the counselor and the client must establish contact, define together "where the client is" in his or her life, and identify the client's current difficulties. This is followed by conversation that leads to a deeper understanding of the client's needs and desires in the context of his or her interpersonal world and to a mutually acceptable diagnosis of the problems. Finally, goals for change are defined, and appropriate actions are planned to accomplish the identified goals. If a client comes to a counselor to discuss a concern that is fairly specific and compartmentalized (such as which of two job offers to accept), the entire sequence of stages may be accomplished in a single session. If a client comes to a counselor with a fairly broad-based or long-standing concern (such as learning how to live as a single parent or dealing with memories of sexual abuse in childhood), the stages may be accomplished over many sessions. Once rapport has been established and in-depth exploration has been undertaken, each problem will be defined. Goals for resolving the problem will then be developed, and, finally, a plan of action for change will be agreed upon and carried out. If new information emerges that changes either the diagnosis or the goals for counseling, the process is adapted to meet these new circumstances.

The First Stage: Initial Disclosure

At the beginning of counseling, the counselor and the client typically do not know one another well. Perhaps the client has seen the counselor in a community education program, a presentation in a residence hall on campus, or a group guidance session at the high school, but usually there has been no prior direct contact. Maybe the counselor has some basic information about the client collected from an intake form or a school record. Neither participant can know in advance the direction their discussion will ultimately take, and the client is probably a bit anxious about disclosing concerns because he or she is not sure how the counselor will receive the disclosures. Cormier and Hackney (1993) describe two sets of feelings clients have at the beginning of counseling: "I know I need help" and "I wish I weren't here" (p. 23). Their description captures well the fundamental ambivalence clients often feel in their initial encounters with a counselor. One central task of the counselor in the first stage of counseling is to allay the client's fears and encourage self-disclosure. Without honest self-disclosure by the client, counseling is an empty enterprise.

Carkhuff (1973) and Egan (1990) both describe attending as important counselor behavior at the outset of counseling. Attending is simply paying careful attention to the client's words and actions. One demonstrates attending by posture, facial expression, and eye contact. As a part of attending, the counselor observes the client's behavior for indications of content and feeling that may not be included in his or her verbal messages. Signs might include fidgeting, tone of voice, failure to maintain eye contact, and so on. We include attending behavior as a part of the initial disclosure stage of counseling because it begins when the first contact between client and counselor occurs. Being attentive to the client's verbal and nonverbal behaviors is the first process goal that is implemented with each client. (Attending to the client's verbal and nonverbal behaviors remains important throughout the counseling process.)

In the initial disclosure stage of counseling, the client must be helped to articulate his or her personal concerns and to place those concerns in a context so that the counselor can understand the personal meanings and significance the client attaches to them. Older counseling literature described this stage as "definition of the problem," but such terminology fails to describe the essence of the initial disclosure process. If, for example, one concern of a client is an unplanned pregnancy, the meaning of that concern will vary greatly depending on the quality of the relationship with the sexual partner, attitudes about the desirability of parenthood at this time, and views on abortion. To define the problem as unplanned pregnancy is only the first step in learning the meaning of the situation to the particular client.

To encourage disclosure, the counselor must set conditions that promote trust in the client. Carl Rogers (1951) described these trust-promoting conditions as the characteristics of the helping relationship:

1. Empathy—understanding another's experience as if it were your own, without ever losing the "as if" quality
2. Congruence or genuineness—being as you seem to be, consistent over time, dependable in the relationship
3. Positive regard—caring for your client
4. Unconditionality—setting no conditions for your caring (not "I will approve of you if you do what I want")

In order to communicate these conditions to the client, the counselor must learn to respond meaningfully to what the client says. At this stage, the most frequent kind of response is restatement, paraphrasing, or interchangeable responding. The counselor keeps the focus of attention on what the client is saying and on the meaning the client attaches to events in his or her life. A typical counselor response might be "You are very angry about the times when your husband goes out and doesn't tell you where he's going or when he will be back." Such a statement tells the client that the content and the feeling of her statement have been heard. If said with appropriate tone, it will communicate caring and genuineness.

Egan (1990) adds another condition that has relevance throughout the counseling process:

5. Concreteness—using clear language to describe the client's life situation

It is the counselor's task to sort out ambiguous statements and help the client find descriptions that will accurately portray what is happening in his or her life. Concreteness promotes clearer insight by the client into his or her life and provides the counselor with a fuller sense of the uniqueness of the client's experience. The following example contrasts a concrete counselor response to a client statement with a vague one.

> CLIENT: I am tired of living this way, but I feel as though there is no way I can get out of caring for my aging father. It all feels pretty hopeless and like no one cares about my needs.
>
> COUNSELOR A: You sound depressed by your situation.
>
> COUNSELOR B: The responsibility of caring for your father has worn you out and left you feeling isolated and without alternatives. You seem depressed.

The response of Counselor B is more concrete because it tells the client that her words were heard exactly. The response of Counselor A could have been made to any number of client statements and does not relate in any specific way to the unique concerns of this client.

If the above conditions are present in the initial disclosure stage of counseling, the client will be encouraged to talk freely and to elaborate on his or her concerns. Gelso and Carter (1985) refer to this point in counseling as the establishment of a "working alliance" (p. 161). In the process, the client doesn't simply tell the counselor what the problem is, he or she begins to clarify the dimensions of life concerns. As the client works to try to communicate his or her ideas and feelings to another, he or she also reaches greater personal understanding. (Chapter 3 presents a detailed analysis of the initial disclosure stage of counseling, along with case material for illustration.)

The Second Stage: In-depth Exploration

In the second stage of counseling, the client should reach clearer understandings of his or her life concerns and begin to formulate a new sense of hope and direction. It is a useful rubric to think of emerging goals as the "flip side" of problems. That is, as problems are more clearly understood, the direction in which the client wishes to move also becomes clearer. At this stage, the goals are not well-defined and the means to reach them are still unclear, but a broad outline of the pattern of desired change is emerging.

The process that facilitates formulation of a new sense of direction builds on the conditions of the initial disclosure stage and becomes possible only if trust has been built in that first stage and is maintained. But the relationship has become less tenuous and fragile than it was at the beginning, and so the counselor can use a broader range of intervention tools without increasing tension beyond tolerable limits. The first stage merges into the second stage as the client's readiness is perceived by the counselor.

In the second stage, the counselor begins, subtly at first, to bring into the discussion his or her diagnostic impressions of the client's dynamics and coping behaviors. The empathic responses of the counselor now include material from prior sessions and focus more on the client's awareness of the unsatisfying nature of old ways of thinking and responding. Such advanced-level empathy statements reassure the client that the counselor has an understanding of his or her world and provide an impetus for still deeper exploration. For example, in the case study earlier in this chapter, Thad, who was in conflict about his responsibilities to his parents, was helped by statements that focused on his apparently conflicting desires for independence and dependence. His problem was a special case of a classic young adult struggle, and the counselor was able to infuse the conversation with a high level of empathy because she knew how important such struggles are for many young people.

As the relationship becomes more secure, the counselor also begins to confront the client with observations about his or her goals and behavior. In the case of Thad, the following confrontative statement was made: "You say you want to be independent, yet you stop at home every evening and end up staying there all evening when you really want to be with friends. Do you think you will establish the kind of independence you need that way?" Broadly speaking, constructive confrontation provides the client with an external view of his or her behavior, based on the counselor's observations. The client is free to accept, reject, or modify the counselor's impression. In the process of considering how to use the counselor's statement, the client arrives at newly challenged and refined views of self. (Many other variations on the theme of constructive confrontation will be offered in Chapter 4.)

Immediacy is another quality of the counselor's behavior that becomes important in the second stage of counseling. The counselor tells the client some of his or her immediate reactions to the client's statements. Immediacy responses often begin with the word *I* rather than *you* so as to identify the content with the counselor, not the client ("I feel angry when you let yourself be mistreated," not "You make me angry when you don't stand up for yourself"). Immediacy responses can be openly supportive ("I am excited about your feeling of success") or confrontative. In immediacy responses that are confrontative, the counselor monitors the client's behavior within the counseling session in order to understand how the client characteristically deals with other people and then shares some of those observations with the client. Here is one example of such an immediacy response: "You seem to be avoiding a decision and acting helpless. When you do this, I have a tendency to want to make your decisions for you." If the client then affirms that this seems to be an accurate observation, it might be followed with a confrontative lead such as "Do you suppose this is what you do with your father, even though you say you wish he would stop trying to tell you what to do?" Immediacy responses work best when the counseling relationship has developed enough that the client is unlikely to interpret the statements as overly critical or unduly supportive.

Since it has been established that the focus of counseling is clearly on the client by the second stage, the counselor may begin sharing bits of his or her own experience with the client without fear of appearing to oversimplify the client's problems or seeming to tell the client "Do as I did." Incidents in the counselor's life may be shared if they have direct relevance to the client's concern. Such self-disclosure can help to establish identity between counselor and client and suggest to the client that he or she is not alone in facing a particular concern. Although some information about how the counselor coped with a similar situation might be relevant to the client's solution, care must be exercised to look for the differences in the client's situation and to permit the client to use the counselor's experience only if he or she sees clear application.

The second stage of counseling frequently becomes emotionally stressful, because the client must repeatedly face the inadequacy of habitual behaviors and must begin to give up the familiar for the unfamiliar. This stressful task must be accomplished within a caring relationship in which it is clear that the counselor is not criticizing the client's past behavior. The thrust is toward helping the client to realize more clearly what he or she does not like in his or her responses to present situations or decision making and to gain a sense of what kinds of responses might be more satisfying.

In the second stage, the counselor and client come to a mutually acceptable diagnosis of the problem or problems. Diagnosis is a process of information gathering and hypothesis testing that results in a description of the problem(s) that takes into account the client's history, life circumstances, and strengths. The diagnosis is determined through mutual discussion of the issues and often through the use of standardized tests that focus on academic, career, or personality variables. Once a diagnosis is established, the counselor and client can move on to the third stage, the identification of specific goals for change and the selection of action plans to implement those goals.

The Third Stage: Commitment to Action

In the third and final stage of counseling, the client must decide how to accomplish any goals that have emerged during the previous two stages. Concerns have been defined and clarified within the context of the client's life situation. The client has considered how his or her own behavior relates to accomplishing the goals that have been clarified through the counseling process. What remains is to decide what, if any, overt actions the client might take to alleviate his or her concerns. If no action is indicated, then the third stage of counseling can focus on increasing the client's commitment to a view that he or she has done everything possible or desirable in the given situation.

The third stage includes identifying possible alternative courses of action (or decisions) the client might choose and assessing each of these in terms of the likelihood of outcomes. Ideally, various courses of action are developed by the client with encouragement from the counselor, although it is acceptable under most circumstances for the counselor to suggest possibilities the client may have overlooked. Possible courses of action and the related outcomes are evaluated in terms of the goals the client wants to attain and the client's value system. Once an action decision is made, the client usually tries some new behaviors while remaining in touch with the counselor. Together, counselor and client monitor the initial steps of the change process. Often the client needs to be reinforced to behave in new ways, both because the old behaviors are habitual and because new behaviors may not bring about immediate results. Particularly when the goals involve improving interpersonal relationships with one or more people, the other parties usually do not respond instantly to the client's new directions, and this can be discouraging. If the client decides that no new action is needed, the decision may be that "I don't need to let myself get so upset by the behavior of another." In such an instance, the reinforcement process supports the ability of the client to manage emotions better when "red flag" experiences occur.

To summarize, the third stage is a decision-making and action time. The client considers possible actions and then chooses some to try out. The counselor gives support for trying new behaviors and helps the client evaluate the effectiveness of new behaviors or new conceptions of reality as they may relate to the reduction of stress. When the client is satisfied that the new behaviors or the new constructs are working satisfactorily, counseling is finished. (See Chapter 7 for more detail on commitment to action.)

THE THREE STAGES OF COUNSELING IN PERSPECTIVE

Table 2.1 provides an overview of the helping process by showing the client's and the counselor's tasks associated with each stage of the experience. Common to all the stages is the recognition that the experience involves work for both participants. Applied to counseling, *work* refers to the experience of exploring with an effort toward understanding more deeply, clarifying what is vague, discovering new insights that relate to one's concerns, and developing action plans.

In the initial disclosure stage, the work of the client involves taking the risk of disclosing information to a relative stranger. It includes an initial effort to make contact with personal beliefs, emotions, and patterns of behavior related to one's issues and concerns. For the counselor, the work of this stage

Table 2.1 Stages in the Counseling Process

	Initial Disclosure	*In-depth Exploration*	*Commitment to Action*
Client work	Communicating the nature of concerns, including content, affect, and context. Clarifying spontaneous meanings of the concerns through the disclosure	Building deeper understanding of the meanings of personal concerns and joining with the counselor to diagnose the major problem(s)	Developing specific goals for change, mustering commitment to reach those goals, and carrying out actions that will accomplish those goals
Counselor work	Providing therapeutic conditions that will build a trusting and working relationship: Empathic understanding Congruence or genuineness Positive regard Unconditional acceptance Concreteness	Extending the client's ability to understand self and others through advanced empathy, immediacy, confrontation, interpretation, and role playing; engaging the client to develop a mutually agreeable diagnosis of the problem(s)	Helping specify goals that may be too general and translating goals into concrete plans for change; re-evaluating unsuccessful actions and rewarding client successes; building a positive end to counseling when goals are achieved

involves developing trust, establishing the counseling setting as a place and time to work, attending intensely so as to understand the significant themes and issues that will need deeper exploration in the next stage, and making initial contact with themes projected in a client's communications. Usually the counselor's work also includes assessing the client's level of readiness for the second stage.

In the in-depth exploration stage, the client's work involves making deeper contact with themes and issues presented in the first stage. It includes clarifying goals for counseling outcomes and developing new discoveries and insights about self and others that relate to these goals. For the counselor, work at this stage involves helping the client develop new understandings by using an artistic combination of advanced-level empathy, immediacy, confrontation, interpretation, role playing, and other structured interventions. Together the counselor and client develop a comprehensive diagnosis of the problem(s).

The client's work in the third stage involves synthesizing new concepts learned in the second stage, developing specific outcome goals derived from the diagnosis, and formulating alternative courses of action. Specific courses of action must be reality-tested, and some are acted upon. Since the action generally involves doing something new, the work also involves gaining control of anxiety that relates to newness. For the counselor, the work involves helping the client synthesize information, formulate goals, and make choices without controlling or imposing self onto the client.

All of these functions will be described in depth in Chapters 3 through 7. We have emphasized that counseling is not simply talk, but work that requires intense energy from both participants. However, it is also true that counseling can be fun and exciting. Saying that counseling is work does not mean that it must be laborious.

Nonlinearity of Stages

As in all stage models, the process described above depicts each succeeding stage as dependent upon the preceding one. That is, unless problems have been defined, it is not possible to establish goals; unless one has established goals, it is not possible to effectively evaluate possible courses of action. Such a linear conception of stages is, nevertheless, an oversimplification, and it is important to recognize the limitations of such a stage model.

First, it is not necessary for a client to clarify all concerns that he or she may have before beginning to think about goals and actions with respect to a particular segment of his or her life. An unemployed head of a household will probably want to work very quickly on securing a source of income. Even though ultimately he or she may want to become involved in an extensive career-planning process, the first question might be "What can I do right now to meet my needs and the needs of my family?" A woman who is a victim of domestic violence may eventually want to consider how she can build a life that does not include her present mate, but first she will want strategies for preventing further violence. Furthermore, a client will typically clarify his or her thinking about some goals at the same time as other goals are in the process of formation. Actions may be taken with respect to the clarified goals, but the client must wait to take action on goals that are still undefined.

Particular actions cannot be evaluated for a goal that has not been defined, and a goal cannot be defined if a concern has not been explored and clarified. Even so, the segments of an individual's life cannot be fully separated and treated as independent problems. Eventually each sector must fit back into a whole picture of the individual's life, much as the pieces of a jigsaw puzzle fit together to produce a complete picture. The process of counseling may involve refining the edges of one piece so that it fits the picture, in which case counseling will take a relatively short time. It may involve the definition of many pieces and work being conducted simultaneously on two or more of them, in which case counseling will last longer. Some pieces may require reprocessing as others take shape. In such a complex case, the process of initial disclosure, in-depth exploration, diagnosis, and commitment to action occurs with each part of the problem. Of course, counseling never returns to the absolute beginning as new sectors of concern are identified, but totally new disclosures may require extensive exploration and clarification before action plans can be considered.

The counseling process also needs to be modified somewhat in certain situations, and these adaptations are reviewed in Chapters 8 through 11. The special needs of female clients, clients in crisis, those who are legally minors, and those whose cultural backgrounds are different from that of their counselor require that counselors be flexible in their conceptualization of counseling and skilled in using a variety of methods to help clients. The ways in which the varying theoretical orientations to counseling affect the process and the ethical and legal duties facing counselors are described in the final two chapters of the book.

SUMMARY

This chapter began with a comprehensive definition of counseling that introduced the broad goal of change in the client, facilitated by discussion with the counselor. Building on the definition, both outcome and process goals were discussed. The client may change behavior, ways of thinking about self and others, or ways of coping with life and making decisions.

The stages of the counseling process were introduced as initial disclosure, in-depth exploration, and commitment to action. The flow moves from the initial meeting of counselor and client, to thorough exploration of client concerns, to goal setting and action planning. Counseling techniques appropriate to each of the three stages were briefly identified. This introduction provides a background for the next five chapters, which contain detailed descriptions of the stages.

REFERENCES

Beck, A. T. (1976). *Cognitive therapy and emotional disorders.* New York: International Universities Press.

Beck, A. T., Rush, A. J., & Emery, G. (1979). *Cognitive therapy of depression.* New York: Guilford.

Brammer, L. M., Shostrum, E. L., & Abrego, P. J. (1989). *Therapeutic psychology: Fundamentals of counseling and psychotherapy* (5th ed.). Englewood Cliffs, NJ: Prentice-Hall.

Carkhuff, R. R. (1973). *The art of helping.* Amherst, MA: Human Resources Development Press.

Cormier, L. S., & Hackney, H. (1993). *The professional counselor: A process guide to helping* (2nd ed.). Boston: Allyn and Bacon.

Egan, G. (1990). *The skilled helper* (4th ed.). Pacific Grove, CA: Brooks/Cole.

Ellis, A. (1989). Rational-emotive therapy. In R. J. Corsini (Ed.), *Current psychotherapies* (4th ed.). Itasca, IL: Peacock.

Gelso, C. J., & Carter, J. A. (1985). The relationship in counseling and psychotherapy: Components, consequences and theoretical antecedents. *The Counseling Psychologist, 13,* 155–243.

Kelley, G. A. (1955). *The psychology of personal constructs.* New York: Norton.

Meichenbaum, D. (1977). *Cognitive behavior modification: An integrative approach.* New York: Plenum.

Patterson, C. H. (1986). *Theories of counseling and psychotherapy* (4th ed.). New York: Harper & Row.

Rogers, C. R. (1942). *Counseling and psychotherapy.* Boston: Houghton Mifflin.

Rogers, C. R. (1951). *Client-centered therapy.* Boston: Houghton Mifflin.

Yalom, I. D. (1980). *Existential psychotherapy.* New York: Basic Books.

CHAPTER 3

BUILDING THE COUNSELING RELATIONSHIP AND FACILITATING INITIAL DISCLOSURE

Chapter 2 described counseling as a process of person-to-person interaction with the goal of stimulating change in affect, thought, and behavior. Changes are regarded as desirable if they help the client make an important life decision, cope more effectively with incumbent developmental issues or life stresses, achieve a sense of personal effectiveness, or learn to use freedom responsibly. Also described in Chapter 2 were three critical interactive stages of the counseling experience: initial disclosure, in-depth exploration, and commitment to action. This chapter is devoted to a consideration of the important properties and related skills of the first stage.

Effective counseling procedures in the initial-disclosure stage lead to sustained self-disclosure by the client for the following purposes:

- To let the counselor know what has been occurring in the client's life and how the client thinks and feels about those events
- To encourage the client to gain some feeling of relief through the process of talking about her or his problems
- To encourage the client to develop a clearer definition of the concerns and greater understanding about exactly what is disturbing through the process of talking about his or her problems

The first of these purposes is obvious since the counselor has nothing to work with until the client does some sharing. It is important for the counselor to try to understand the client's experience thoroughly and to perceive how the client thinks and feels about what has occurred. Beginning counselors often overlook the importance of the other two purposes. Reflecting on your personal experience, try to recall a time when you talked about an important problem with someone else. People typically feel better after sharing a concern with someone else, even if no new solutions are evident. This process, known as catharsis, is explained by the emotional release that comes from "getting the problem out." Finally, you may have experienced the process of trying to explain your situation to someone else, only to discover that in the telling of your story you reach new understandings of how things fit together. Clients frequently gain such new insights, too, through the process of telling.

Relationship building is the first important step in the counseling process. If the total counseling experience is to be of benefit to the client, time and energy must be devoted to developing a relationship that can be characterized by mutual trust, openness, comfort, and optimism about the value of continued counseling sessions. These supportive conditions provide the necessary basis for counseling to evolve into an experience of deeper exploration, which constitutes the second stage of counseling.

In describing the initial-disclosure stage of counseling, we will first discuss the characteristics that clients bring with them to the first counseling session. Next, characteristics and skills of the counselor that facilitate the relationship and lead to initial disclosure will be described. The effects of initial disclosure for the client will then be examined. Finally, behaviors that are often seen as helpful by well-intentioned people are shown to detract from relationship building and to block exploration in the early phases of counseling.

WHAT CLIENTS BRING TO THE COUNSELING EXPERIENCE

Some clients enter the counseling experience voluntarily. They become aware of tension, anxiety, dissonance, confusion, or lack of closure in their lives, and they realize that they need assistance. The degree of felt tension may vary from mild (as in the case of a high school senior who indicates a desire for more information about college scholarships and loans) to intense (as in the case of a married couple who are experiencing conflict leading to domestic violence). Voluntary, or self-referred, clients have several characteristics in common: they experience tension and conflict, the tension motivates an effort to seek help and to initiate at least some disclosure about themselves and their concerns, and they are willing to do some serious thinking. For such clients, an initial question such as "How can I help?" from the counselor is usually sufficient to initiate relevant communication.

Other clients are reluctant to participate in counseling. They may be experiencing tension but do not acknowledge the need for help. Examples of reluctant clients include individuals who are abusing drugs or alcohol, victims of domestic violence, people whose behaviors are aggressive and disruptive to others (and may have led to arrest or legal problems), and students who are unmotivated and unsuccessful in school. Because such clients are generally pressured (or required) by a third party to make initial contact with the counselor, they often put up strong barriers to genuine participation. They may avoid offering relevant disclosures and doing the work that can lead to growth. Working effectively with reluctant clients requires some special skills, which will be discussed in Chapter 10. In this chapter, we assume that the client has entered counseling with some willingness to initiate disclosure and to do some work.

Even clients who seek counseling voluntarily usually experience some anxiety about the process and some resistance to facing difficult issues. Some of the anxiety stems from the inherent condition of the counseling process, which requires the client to share information about self and personal life with a person who is initially a relative stranger. Critical questions clients ask themselves include: "What will this person do with the information I share with him or her?" and "What impressions of me will this person develop if I honestly describe my concerns, stresses, and doubts?" Some of the anxiety comes from the unknown: "What will happen to me during this experience?"

Since caution and anxiety are usual, the counselor's objectives during the initial-disclosure stage are to help the client feel comfortable about the process of communicating and less anxious about exploring fully, honestly, and in depth those concerns requiring attention. Counselors initiate disclosure with three interconnected patterns: by inviting communication, by responding to client concerns in a caring way, and by avoiding doing and saying things that block communication.

WAYS TO INVITE COMMUNICATION AND BUILD THE COUNSELING RELATIONSHIP

The Counselor's Nonverbal Messages

You have been with people whose body language invites communication and others whose body language indicates disinterest and perhaps even anxiety about communicating. The active, interested listener faces and leans toward the communicator in a posture of interest and even excitement. Eyes are focused in the general direction of the communicator's face. Arms are in an open mode in relation to the trunk, as if to say "I am very interested in receiving, with all my sensory processes, what it is you want to say to me." The attentive listener maintains an interested facial expression and makes encouraging gestures (nods, smiles, hand gestures, etc.). The energy for offering these cues seems to come naturally and without contrived effort; it is a bodily representation of a strongly held belief about how to receive and welcome another human being.

These skills of reception are referred to as attending skills, and quite literally they communicate that the counselor's undivided attention is on the client's concerns. Egan (1990) defines attending as "the way you orient yourself physically and psychologically to the client" that results in "a certain intensity of presence" (p. 108). Attending behavior encourages talking by the client and therefore reduces the need for the counselor to talk (Ivey, 1988), placing the content of the session more in the client's control.

In contrast, the listener who is disinterested and uncomfortable does not use his or her body to invite communication. Facial focus may be forty-five degrees or more away from the communicator's face, with arms folded in a self-protective position, eyes on the desktop. Body cues like these communicate messages such as "I'm not interested," "I don't care," "I don't have the energy to be available to you," and "I'm afraid to be open with you." Cues that indicate that the listener is closed can be subtle, often outside the person's awareness.

Habits of attending vary somewhat from one culture to another. For example, the patterns of eye contact that have been described here are typical for whites in the North American culture. Ivey (1988) indicates that African Americans often show a pattern of greater eye contact when speaking than when listening and that Native Americans may avoid eye contact altogether when speaking of important matters. The skilled counselor learns to respect these cultural differences.

Ivey (1988) also suggests that the counselor can shape what the client talks about by intentionally showing greater attentiveness when the client is discussing pertinent material and less attentiveness to nonproductive talk. In the initial stages of a relationship, however, intentional withholding of attention might easily be misinterpreted by the client as a lack of caring. Furthermore, a counselor may inadvertently limit talk about material that is important to the client but that makes the counselor feel uncomfortable. Ivey uses the example of how eye contact may be interpreted by the client as permission to talk about sex or as an indication that the topic is inappropriate.

In addition to the nonverbal messages of body language, the counselor also communicates attending or caring by means of voice characteristics and manner of speaking. These qualities are sometimes referred to as "paralinguistic," because they are communicated by voice but have nothing to do with verbal content. For example, rapid speech, stumbling over words, loud tone, or tightness of voice may signal stress. Slow, quiet, listless speech may indicate inattentiveness. Vocal tone and emphasis can highlight certain material as important. Comfort may be communicated by a pleasant tone, purposeful but comfortable pacing, and other qualities of voice.

Finally, it is important to realize that just as the counselor communicates comfort and attention by his or her nonverbal behaviors, the client also gives many cues to his or her emotional state through these behaviors. Brammer, Shostrum, and Abrego (1989) suggest that the nonverbal behavior of the counselor should match the nonverbal behavior of the client—mirroring, for example, body posture, position, breathing, voice volume, rate of speech, and other qualities.

Though many of the attending skills seem commonsensical, it is nevertheless true that persons in positions to be helpful frequently do not exhibit these behaviors. Can you think of examples of persons who were deficient in attending skills and thus made necessary communication difficult? Gilliland and James (1993) state that attending is both an attitude and a skill and that practice leads to improved performance.

Exercise

Form groups of three students: one will be the helpee, one will be the helper, and one will observe. The helpee should think of some issue to discuss with the helper. The helper should intentionally violate the principles of good attending described above as the helpee presents his or her issue. The observer notes whatever occurs. Continue this pattern for three minutes. Then discuss what the helper's behavior did to the process. Did the helpee show any nonverbal signs of his or her reactions? The helpee can share directly how it felt to be ignored while trying to talk. If this exercise is done in a classroom setting, debrief the various triads and share experiences across groups.

Repeat the same process (switching roles so that different participants serve as helper, helpee, and observer) for an additional three minutes, with the helper practicing the best attending skills possible. Discuss the contrasts with the first session.

Finally (after switching roles again), a helper and a helpee should engage in a discussion about something important to the helpee for a longer period, perhaps seven minutes. The helper should maintain good attending behavior and may offer brief verbal responses as appropriate. The observer will take written notes on the paralinguistic and attending behaviors of both the helper and the helpee. At the conclusion of this session, the observer shares his or her findings, and the members of the triad discuss the meanings they attach to the behaviors that were observed. Both the helper and the helpee are encouraged to be self-disclosing about how they felt when a particular behavior occurred in the session.

The Counselor's Verbal Messages

Encouraging of communication usually begins with the counselor offering an honest invitation to communicate. A statement such as one of the following is usually sufficient to offer an invitation: "How can I help?" "What would you like to discuss today?" "How would you like to begin?"

Most self-referred clients will respond to such invitations with an expression of concern accompanied by an implied need for help. Initial client statements that we have heard recently include:

- My husband is an alcoholic and has been that way since before we were married. I can't get him to stop and I'm scared and I don't know what to do.
- I find that I am more sexually attracted to men than to women, but I really feel guilty because I know it's wrong.
- Tommy keeps bothering me on the playground. He pushed me off the jungle gym again, and yesterday he punched me.
- It's just two weeks till graduation, and I don't have any idea what I'm going to do next.
- Well, my daughter really seems to be out of control. She stays out till all hours of the night and refuses to talk about what she's doing.

Such statements express an initial concern. Further development and clarification are always necessary. Follow-up invitations that the counselor can use to encourage further development include such statements as these:

- Tell me more about . . .
- Help me understand more fully . . .
- Tell me what happened when . . .
- Help me understand your thinking about . . .
- It sounds as though you feel . . .

THE CORE CONDITIONS OF COUNSELING

The preceding section described some specific counselor behaviors, both nonverbal and verbal, that invite a client to begin talking about his or her concerns. Many beginning counselors are able to accomplish these first steps with ease, but then have difficulty knowing what to say when the client begins to disclose. It is helpful to remember the purposes of disclosure. The effort is to get the client to share his or her experience, to release pent-up feeling in the process of telling the story, and to begin to clarify the true nature of the problem or problems presented. The counselor's goal is to come to understand the client's experience as clearly and as personally as possible.

New counselors are often so predisposed to finding a solution to the problem(s) that they forget that it is ultimately the *client's* responsibility to solve the problem(s). Suggestions made by counselors for the resolution of clients' concerns prior to careful disclosure short-circuit the counseling process, demean the client's ability to be self-directing, and often provide solutions that don't fit the situations.

To support clients' disclosure of meaningful issues during the initial-disclosure stage of counseling, the counselor maintains an attitude of receiving the client, often referred to as the core conditions of counseling. Three of these conditions—empathy, positive regard, and genuineness—were described by Carl Rogers (1957) as the necessary and sufficient conditions of therapeutic personality change. The fourth condition—concreteness—is the counselor's skill in focusing the client's discussion on specific events, thoughts, and feelings that matter, while avoiding a lot of intellectualized storytelling. Concreteness is a precaution against the rambling that can occur when the other three conditions are employed without sufficient attention to identifying the client's themes. Concreteness will be evident to the degree that the counselor identifies and responds to important client themes, while choosing not to respond to small talk, excessive storytelling, and other material of a social or diversionary nature.

Empathy

Rogers (1961) defined empathy as the counselor's ability "to enter the client's phenomenal world—to experience the client's world as if it were [the counselor's] own without ever losing the 'as if' quality" (p. 284). Subsequent authors (Carkhuff, 1969; Egan, 1990; Ivey, 1988; Patterson & Eisenberg, 1983) have reasoned that there are different levels of empathy and that, to stimulate client exploration, the level of empathy communicated must be related to the client's level of readiness for disclosure. Primary empathy is the level that is facilitative in the initial-disclosure stage of counseling, and advanced empathy is appropriate for the in-depth exploration stage (to be discussed in the next chapter). This section develops some overall understandings about empathy and focuses on the primary level.

Empathy involves two major skills: perceiving and communicating. *Perceiving* involves an intense process of actively listening for themes, issues, personal constructs, and emotions. *Themes* may be thought of as recurring patterns—for example, views of self, attitudes toward others, consistent interpersonal relationships, fear of failure, and search for personal power. *Issues* are questions of conflict with which the client is struggling: "'What do I want for my future?" "How can I be sexually active and 'safe'?" "Why does every event in my family turn into a disaster?" "Why do I still feel that I am fat, when I have lost a lot of weight and am in fact thin?" Relative to each theme or issue, a client will have emotions of elation, anger, anxiety, sadness, confusion, etc. Understanding the emotional investments is a critical part of the perceptual element of empathy.

George Kelly (1955) described the perceptual element of empathy as understanding the client's personal constructs. He defined *personal constructs* as the unique set of thoughts a person uses to process information, give meaning to life events, order his or her world, explain cause-effect relationships, and make decisions. Personal constructs include beliefs about self and others, assumptions about how and why events happen in the world, and private logic and moral premises that guide the person's world. The counselor can detect the client's personal constructs from his or her descriptions of life situations, behaviors, decisions, or even responses to tests and inventories.

Beck (1976) and Meichenbaum (1977) discuss "automatic thoughts" and "internal dialogue," respectively. Each describes how the client's cognitive structure (complex of personal constructs) predisposes him or her to interpret events in certain ways. Emotions emanate from these interpretations. Thus, empathy includes knowing not only the events that have occurred in the client's life but also how his or her cognitive structure has led to interpretation of those events and to consequent feelings.

As the client discloses information about self, some of the client's themes form very quickly in the counselor's perceptual foreground while others remain in the background. Some personal constructs (for example, "The world is not a safe place" or "People are phony") may emerge quickly, although others will remain unavailable. Often as a counselor listens to a client's story, the counselor will see errors in the client's logic, construct cause-effect relationships that differ from the client's, and identify the basis of the client's distressed feelings. This diagnostic thinking guides treatment planning and serves as a basis for the counselor's responses later in the counseling process.

In the communication component of empathy, the counselor says something that tells the client that his or her meanings and feelings have been understood. If a counselor listens carefully and understands well but says nothing, the client has no way of knowing what is in the counselor's mind. Sometimes a client may even misinterpret a counselor's lack of response as a negative judgment about what has been said. It is often through hearing his or her meanings and feelings repeated that a client takes another look at life events and begins to perceive them differently.

Primary empathy is most often communicated through an interchangeable verbal response (though facial expressions and other nonverbal responses can also be used). Interchangeable responses are statements that capture the essential themes in a client's statement but do not go deeper than the transparent material. A paraphrase such as "You felt degraded and angry because your girlfriend criticized you in front of your friends" is a fairly typical response of this type. It captures both the feeling and the meaning of the client's disclosure in simple language that the client can understand.

Feeling understood, the client will very likely continue to elaborate on the meaning of that or related experiences. It is important to realize that statements such as "I know just how you felt" do *not* communicate empathy because they contain nothing of what the client has shared. Advanced empathy (to be discussed more fully in Chapter 4) is communicated through additive verbal responses, in which the counselor adds perceptions that were implied but not directly stated by the client. Ability to hear these implied meanings grows with experience and with the quality of the counselor's diagnostic thinking.

In the first stage of counseling, primary empathy is used more because it demonstrates to the client that the counselor is listening effectively, without the threat that can occur if the counselor seems to be seeing through defensive strategies too quickly. The client's sense of progress and of comfort are best served at this relationship-building stage if the counselor is seen as perceptive—but not too perceptive.

A good example of an interchangeable response comes from a classic counselor training film in which Carl Rogers (1965) is counseling Gloria, a divorced client with two young children. At one point, Gloria says to Rogers, "I really don't know how my daughter [who was 7 years old] will feel about me if she knows that I sleep with a man I'm not married to." Rogers restates Gloria's concern: "If your daughter knew about your actions, would she, *could she,* accept you?" Gloria affirms Rogers' understanding and goes on to describe her anxieties about being a good mother and being accepted by her daughter. Notice that Rogers identified a key theme and reflected it back to Gloria in a way that helped her look at it more deeply. He did not simply repeat the statement but focused it without adding anything to it.

Cultural sensitivity and knowledge of cultures different from one's own are important to the effective use of empathy. Okun (1992) cautions, "While there are some basic skills and strategies that cut across class, race, and culture, helpers must adapt their counseling style to achieve congruence with value systems of culturally diverse clients" (p. 8). Cultural background influences not only the personal constructs through which an individual interprets the world but also the style of expression that is experienced as empathic. For example, Ivey (1988) indicates that verbal responses that move too quickly to meanings and feelings are often considered to be rude in Asian cultures.

Effectively communicated empathy has a number of desired effects in the initial-disclosure stage of counseling. First, the energy required to listen actively expresses caring and affirmation to the client. The counselor is saying, "I care enough for you that I want to invest energy into understanding clearly."

Second, the feedback that comes from the counselor's contact with significant themes helps the client see his or her own themes more clearly. This helps the client understand self more deeply and reexamine relevant perceptions, attitudes, and beliefs.

Third, such responding establishes expectations about the nature of the counseling experience. Counseling is conveyed to the client as a process that involves attending to self, exploring, searching, and perceiving self more clearly. Counseling is established as an experience involving work, not simply conversation. Indeed, the work of the counselor is to stimulate the client's work of self-discovery.

A fourth effect is that a client will feel that it is safe to continue the counseling experience if a counselor is careful to offer a level of empathy that is consistent with the client's level of readiness. The client learns that nothing bad will happen as a result of communicating and that something helpful is likely to occur.

A fifth effect is that empathy communicates to the client that the counselor has special expertise to offer. Empathy is not routinely experienced in the events of daily life. A counselor who can make empathic contact establishes himself or herself as having some special skill. In turn, this helps the client experience a sense of optimism about future sessions.

Positive Regard

Positive regard is caring for clients for no other reason than the fact that they are human and therefore worthy. Caring is expressed by the enthusiasm one person shows for being in the presence of another and by the amount of time and energy one is willing to devote to another's well-being. The experience of being cared about helps develop and restore a sense of caring for self. It creates energy and encourages a person to respond to the demands of life. A counselor's caring can increase a client's enthusiasm for work and growth.

Rogers (1957) developed the concept that the counselor's caring for the client can be unconditional. Since the counselor does not have a role in the client's life outside the counseling situation, he or she can become the client's instrument for change without holding a lot of preconceived ideas about what behaviors the client should exhibit. The counselor's respect for the dignity and worth of the individual remains intact regardless of the client's behaviors. Parental love—which is probably the most important support a growing human ever experiences—is very much like unconditional regard, but parents can never fully achieve unconditionality because they have a vested interest in what their children do. They believe that their children's behavior reflects on their effectiveness as parents and by extension on their worth as individuals. As a paid professional helper, a counselor does not have a comparable ego investment in his or her clients. It is, of course, an ideal that the counselor can care for (love) each client equally. All counselors, being human, meet clients who are hard to like. It is important when this occurs for the counselor to examine and work through his or her feeling of disregard or to refer the client to another counselor. If a counselor does not have a feeling of positive regard for a client, one of the necessary conditions for therapeutic change is missing.

To work through feelings of disregard for a client, a counselor must first acknowledge them and take responsibility for their existence. After recognition, the counselor's task is to identify specific characteristics of the client that he or she does not like. For many counselors, lying, defensiveness, manipulation, destructiveness to self and others, unwillingness to conform to reasonable social rules, and irresponsibility toward others are traits that often trigger dislike. For example, some counselors might have difficulty working with a client who is known to have been the perpetrator of sexual abuse on a child. After taking whatever steps are necessary to secure the safety of the victim, the counselor's task is to become committed to helping the client get beyond this very serious problem in his or her life. Though the client's behavior may be repugnant if judged through the moral imperatives by which the counselor lives his or her own life, the counselor tries to understand the meaning of the behavior in the client's life without judgment so that strategies for change can be devised.

Several principles of human behavior may help counselors work through their emotions. One is that a counselor may be tempted to impose "should" statements on a client. In the case of a client who is an unfaithful marriage partner, the counselor may feel that the client should behave more responsibly toward his or her spouse. In the case of adolescents who are skipping school, many would say that they should attend. But a counselor attempting to *impose* either of these "shoulds" would be experienced by the client as uncaring. The likelihood is that the client already understands that there are negative consequences to his or her behaviors but has still chosen those behaviors. Rather than being rejected for those behaviors, he or she needs help in finding alternative behaviors that will have more positive consequences.

A second principle is that anxiety often accompanies feelings of dislike for a client. A counselor may feel threatened by client behavior that arouses his or her own unresolved issues or by fear that the client's problems are beyond his or her ability to help. Excessive resistance by the client or power struggles in the counseling sessions can also trigger counselor anxiety.

A third principle is that some characteristics of a client may remind a counselor of some other person for whom there are feelings of anger or resentment. In such circumstances, the counselor does not perceive the client with full accuracy but instead has some distortions in his or her image of the client. (See the discussion of this phenomenon, called countertransference, in Chapter 10.)

Based on these principles, working through dislike for a client requires that the counselor give honest answers to the following questions:

- What characteristics of my client do I find unlikable?
- What do I think my client should be doing that he or she is not?
- How are my "shoulds" affecting our relationship and my openness with the client?
- If I am imposing "shoulds," what am I missing about my client as a result?
- Am I experiencing an anxious rather than a calm feeling with my client?
- With whom in my life might I have important unfinished business?
- Is my own unfinished business interfering with my ability to feel caring for this client?

These questions are very difficult for any counselor to answer without the assistance of a professional colleague. For this reason, we strongly recommend that every counselor have his or her own resource counselor—usually either a colleague or a supervisor. Exploration of feelings of disregard for a client with another counselor will help a counselor determine whether the disregard can be understood and replaced with a more helpful attitude or the client should be referred. The counselor may also learn that his or her own unresolved issues are intrusive in enough counseling interactions that personal therapy may be needed.

Of course, an effective counselor will experience positive regard for the vast majority of his or her clients. Although caring is usually not as directly expressed as empathy, it will become apparent to the client through the counselor's spontaneous statements that acknowledge the validity of the client's struggle for a more satisfying life. As Ivey (1988) states, "the counselor points out how, even in the most difficult situation, the client is doing something positive" (p. 130). Similarly, a client will usually detect the absence of caring fairly quickly, so the counselor is well advised to deal promptly with such feelings.

Genuineness

Rogers (1942) originally defined genuineness as the characteristic of transparency, realness, honesty, or authenticity. He has also used the term *congruence* to suggest that a genuine counselor behaves in ways that are congruent with his or her self-concept and thus consistent across time. The counselor shares thoughts and feelings in ways that do not manipulate or control the client. Although genuineness does not give the counselor license to ventilate his or her own emotions on the client, Rogers said that a counselor who is having trouble liking a client would do well to share and try to resolve the feeling with the client rather than trying to hide it. We caution that clients with deep wounds to their self-esteem may not be able to handle such honesty from a counselor. Genuineness may have to be established through persistent congruence over a longer period of time.

Rogers believed that if the counselor behaves consistently over time, he or she eventually will be perceived as real. Having this perception of the counselor will help the client feel safer and more trustful and thus more willing to engage in intensive self-exploration. Experiencing genuineness from the counselor in a climate of safety enables the client to be more genuine and also encourages him or her to drop defenses, games, and manipulations. Transparency, that is, allowing the client to see into the counselor's thoughts and feelings, reduces the client's concern that there are hidden agendas—that the counselor is going to try to manipulate him or her into behaving in certain ways.

To be fully genuine in the sense described by Rogers, the counselor must know himself or herself very well. He or she must have a clear picture of his or her personality and how these characteristics are expressed in significant events and relations with people.

The principle of genuineness dictates that a counselor never communicate dishonestly, never present information that misleads, and never knowingly present an image of self that deceives a client. However, the principle does not require that the counselor impulsively disclose every thought, opinion, and feeling to a client. Sharing information about self is a decision, not an impulse, for both counselor and client. The counselor decides what and when to share based on perceptions of the client's need for or ability to benefit from the information. For example, sharing an emotion experienced in the client's presence or an observation about what is happening right now in the relationship (such as "I feel that you are unloading on me today") is immediacy communication that the client may have trouble working with during the initial-disclosure stage. In the next stage, such responding may be appropriate because the client feels safer and is more ready to work with it. Disclosing a past experience that parallels the client's experience may help to reduce distance and create a feeling of mutuality. However, going into a lot of detail about past personal experiences can quickly lead to storytelling that can take the focus away from the client and block the exploration process. If the counselor chooses to disclose personal experiences, he or she should disclose no more than is necessary to help the client see the parallels with his or her own experience.

Another type of disclosure that tends to diminish the client's perception of genuineness is opinions about other people's behavior. Clients who experience interpersonal stress sometimes want confirmation that they are right and others are wrong. They may try to draw the counselor into taking a position. For example, with couples experiencing marital disagreement, each client may seek to gain the counselor as an ally against the other. Since the counselor cannot know all the circumstances or motives involved and since talking about another person's behavior is gossiping, it is wise for the counselor not to express any judgment. A counselor who makes the error of judging a third party's behavior will be seen as a counselor who will judge the client as well. It may help if the counselor explains that he or she is not in a position to judge.

Exercise

Think of two people in your life. Person A is an individual you perceive to be genuine. Person B is an individual you do not perceive to be genuine. Develop a clear visual image of each person. Recall one or two significant experiences that you have had with each one. Now, while remembering these experiences, answer the questions below. Write down your answers or share them with another student.

1. What specific observations have I made about person A that give me the impression that he or she is genuine?
2. What specific observations have I made about person B that give me the impression that he or she is not genuine?
3. What differences do I note in their ways of relating to me?
4. How would I describe my inner experiencing in the presence of person A, particularly my emotions?
5. How would I describe my inner experiencing in the presence of person B, particularly my emotions?
6. From my personal experience, what principles about genuineness seem valid for me?

Concreteness

As Ivey (1988) has stated, "a concrete counselor promptly seeks specifics rather than vague generalities. As interviewers, we are most often interested in specific feelings, specific thoughts, and specific examples of actions" (p. 130). Concreteness is *not* one of the Rogerian conditions of the helping relationship. In fact, the concept has emerged because it has been observed that a counselor who is empathic, caring, and genuine may still encourage a client to ramble a great deal and allow the client to avoid talking about important material. If the counselor responds with equal interest to the statements "The weather has been very nice" and "The way my husband has been treating me and the kids really sucks," the client is likely to be encouraged to elaborate on either statement as though they were of equal importance.

As stated earlier in this chapter, it is the counselor's responsibility to identify which of the client's statements are central to his or her reasons for being a client and to encourage talk about those issues. The client is still the person who determines what will be introduced as the content of the session, but the counselor manages the process in such a way as to make it easier for the client to talk about what matters. What the counselor responds to, the client will probably follow up on; what the counselor ignores is likely to be dropped. As a counselor's diagnostic skills improve with experience, it becomes easier for the counselor to identify important themes to be pursued. Even the beginning counselor, however, can easily distinguish between small talk and self-disclosure. Beyond initial social amenities that may contribute to client comfort, small talk wastes valuable counseling time.

Ivey (1988) suggests that concreteness can be increased by asking directly for specific examples of troublesome events. For example, the frequently heard complaint "She is always picking on me" tells little, but examples of specific interactions between the client and the other person will shed much light on the relationship dynamics. "Picking on me" may actually mean "Every time I don't have my homework done, the teacher calls attention to it in a public way and embarrasses me."

The language used by the client and by the counselor can also contribute to unfocused discussion. Vagueness, abstractness, and obscurity are the opposites of concrete communication. Therefore, the counselor should model direct communication as well as challenge the client to become more specific. The more fully and concretely the troublesome events in the client's life are recreated, complete with affective tone, within the counseling session, the more likely it is that new understandings and more positive feelings can be developed.

WAYS TO BLOCK COMMUNICATION

Just as certain attitudes and behaviors can facilitate client self-disclosure, other predispositions and behaviors sabotage communication and build roadblocks to such disclosure. Many of the ineffective responses that new counselors make are a result of carrying certain social behaviors and an "expert" model of helping into counseling sessions. Counselor behaviors such as advice giving, lecturing, excessive questioning, and storytelling make client self-disclosure very difficult.

Counselor Predispositions

People learn to avoid emotional content in conversations and social interactions with others unless the relationship is strong and the circumstances are private. For example, a couple at a party who get into a disagreement over how to manage money will be avoided by the other party guests. People learn how to give other people social distance and privacy. If a counselor employs the same approach, the impact is to communicate to the client that it is not safe to share emotional content. It is important to remember that the client sees the counselor as a stranger in the first session. The goal is to establish trust as quickly as possible by employing the attitudes and behaviors described in the previous sections. Another attitudinal barrier that many counselors struggle with initially is the feeling that as experts they ought to be able to solve clients' problems and relieve them of their psychological pain very quickly. This attitude will short-circuit the disclosure process because the counselor adopts the client's anxiety and tries to move prematurely to solutions.

Premature Advice Giving

"Why don't you"/"Yeah, but" (as described in transactional analysis) is a verbal game frequently played in counseling sessions. A client enters, communicating cues of confusion, stress, exasperation, and helplessness. Wanting to offer some relief, the counselor gives some advice ("Why don't you . . ."). The client does not respond with enthusiasm to the advice ("Yeah, but . . ."). Soon barriers are built in the relationship. The client begins to feel more defensive. Feelings of exasperation grow in the counselor and may result in rejection of the client.

In addition to blocking communication, premature advice giving has the effect of making the client feel like an inadequate, dependent child, and thus it blocks the development of responsible self-reliance. Indeed, clients sometimes present themselves as helpless children in an attempt to seduce counselors into giving advice. Giving advice under these conditions reinforces defensive manipulations and game playing.

A counselor's thoughts about how a problem may be solved are much more appropriate later in the counseling process, after the circumstances have been thoroughly explored and the counselor *really* knows something of the client's experience. At that point, the counselor's ideas can be considered along with the client's as options that might be useful.

Lecturing

Lecturing is really a disguised form of advice giving. When lecturing, a counselor presents himself or herself as an expert, developing a few paragraphs or more of "sage wisdom." This is the hidden message: "You should do it my way. If you don't, you are stupid and I will be angry with you." Disregard for the client is shown in several ways. In addition to conveying a "stupid" message, the counselor is allocating more time to his or her own material than to the client's. During long-winded lectures, clients frequently shut off their listening receptors and often disregard the message giver as well as the message itself. This is especially true for clients who have been involved in power struggles with authority figures. The counselor should not be seen by the client as just one more authority figure. *You know you are probably lecturing if you say more than three consecutive sentences to your client.*

Excessive Questioning

"How old are you?" "Where do you live?" "How many siblings did you grow up with?" "What religion did you affiliate with as a child?" "When did you reach puberty?" "Do you like to be outdoors?" "Do you watch much television?" "Why are you taking the course for which you are reading this book?" "What do you want to do with the rest of your life?"

Excessive questioning does not open communication; it conveys that the client's job is to be *passive* and *respond* to the counselor's questions. The counselor asks and the client responds. The problem is that the client has little opportunity to do any initiating. The counselor is in charge, but the client is the person who knows what the issues that need airing are. Excessive questioning results in important information being missed. (Have you ever been in a situation where you had important information to share with another person, and his or her questions were focusing on something else and limiting your ability to say what you needed to say?)

"Why" questions are especially destructive because they ask a client to explain and justify his or her behavior. They move counseling away from a personalizing experience.

Sometimes counselors ask questions to stimulate interaction with a nonverbal client. Sometimes they ask questions to get information. Sometimes the intent of a question is to direct a client's attention to a specific topic, theme, or issue. All of these goals are valid, but responses in forms other than a question are usually more facilitative for accomplishing these ends in the initial-disclosure stage of counseling.

To help a nonverbal client, a counselor might use alternatives such as "I can see that it is hard for you to talk comfortably with me. Perhaps it would help to share with me something that is happening in your life that creates excitement for you." Notice that such a response works with, rather than against, the client's energy.

If there is information that is important to know, a counselor might try an alternative such as "I need to understand more clearly how things looked, sounded, and felt to you when your wife decided to move out. Help me understand what it was like for you then." Such responses create mutuality in the counseling experience. Before making such a statement, the counselor must first clarify that the information being requested is very important.

To focus a client's attention on a specific theme, an alternative such as the following can be helpful: "You have mentioned sitting alone on a school bus a couple of times. Tell me the things you think about when this happens." Again, such an alternative invites exploration in a comfortable way, while promoting concreteness in the content of the discussion.

Unless it probes too deeply into very private space, an occasional question is unlikely to be damaging to the client-counselor relationship. Questions are detrimental if they become the dominant mode of counselor communication or if they create a stifled rather than a spontaneous atmosphere.

Beginning counselors can control the tendency to regress into questioning by never asking two questions in a row, by following a question with a nonquestion that relates to the information received in response to the first question, and, if necessary, by using a general lead such as "Tell me more about . . ."

It also helps to distinguish between closed and open questions. *Closed questions* usually probe for specific factual information and can be answered with a single word or a few words. *Open questions* call for ideas, beliefs, or emotions and require longer discussion for full response. "Are you living with your husband now?" is a closed question. "What do you do when your husband comes home drunk?" is an open question. Open questions allow for fuller client disclosure and are usually better than closed questions during the first stage of the counseling process.

Storytelling

COUNSELOR: I remember when *I* was in the tenth grade. The kids in my group were really self-centered then, too. Everybody seemed interested in looking out for number one. None of us knew anything about being a friend. Everybody seemed to be putting a lot of energy into building themselves up by putting others down.

CLIENT: Gee, I guess life must have been pretty lonely for you at 15.

COUNSELOR: Yeah, it really was. Why I remember a time when . . .

Counselors often tell stories about their own lives, ostensibly to help their clients identify with them and to communicate this message: "I understand what you are experiencing because I experienced it, too." Generally such storytelling blocks rather than facilitates communication. First, it is often difficult for clients to believe that the counselor could have ever been in a situation exactly like theirs, and usually they are correct in their skepticism. Furthermore, an underlying presumption is that, in a similar situation, the counselor held the same beliefs and experienced the same emotions. By moving the focus of discussion from the client, the counselor interrupts the exploration process. He or she discounts the client and risks introducing "interfering noise" into the session by projecting his or her own emotions onto the client. Storytelling is often a way for the counselor to cope with his or her personal anxiety. At the worst, the counselor may be using the client as an audience to gratify a need for attention.

SUMMARY

The primary task of the initial-disclosure stage is to work toward the development of a relationship characterized by trust, honesty, and open communication and to promote self-examination and clarification. As the experience evolves, the client discloses progressively more information about self and personal concerns. The sharing of this information helps the client gain new insights, helps the counselor understand what is occurring in the client's life, and helps both to work toward tentative goals for counseling.

As the client discloses information, the nature of contact with the counselor gives the client feedback about whether it is safe to continue. This contact also establishes counseling as a process of self-examination as opposed to mere conversation. It should also extinguish any client predisposition to see the process as an interview in which the counselor collects information and then provides a solution.

Effective contact involves encouraging communication with verbal and nonverbal initiatives and communicating empathy appropriate to the client's level of readiness as well as genuine concern about the client's welfare. Empathy can be described as the ability to perceive and understand themes and issues projected by the client's communication. It can also be defined as perceiving and understanding the personal constructs a client uses to understand and organize his or her world. Empathic skill includes a communicating as well as a perceiving component, and the most common way of expressing empathy is through interchangeable responses. Concreteness is a quality by which the counselor keeps the session focused on important themes and promotes clear and specific description of the client's life situation.

Several communication patterns that are often thought to be helpful by new counselors and untrained persons in general are advice giving, lecturing, questioning, and storytelling. In fact, these behaviors usually block communication in the initial-disclosure stage of counseling. Limited use of questions, advice, and self-disclosure (storytelling) may be appropriate and helpful later in the counseling process, but usually not at the beginning.

The development of an effective working relationship during the initial-disclosure stage establishes the foundation necessary for the next stage: in-depth exploration.

REFERENCES

Beck, A. D. (1976). *Cognitive therapy and emotional disorders.* New York: International Universities Press.

Brammer, L. M., Shostrum, E. l., & Abrego, P. J. (1989). *Therapeutic psychology* (5th ed.). Englewood Cliffs, NJ: Prentice-Hall.

Carkhuff, R. R. (1969). *Helping and human relations* (Vols. 1 & 2). New York: Holt, Rinehart and Winston.

Egan, G. (1990). *The skilled helper: A systematic approach to effective helping* (4th ed.). Pacific Grove, CA: Brooks/Cole.

Gilliland, B. E., & James, R. K. (1993). *Crisis intervention strategies* (2nd ed.). Pacific Grove, CA: Brooks/Cole.

Ivey, A. E. (1988). *Intentional interviewing and counseling* (2nd ed.). Pacific Grove, CA: Brooks/Cole.

Kelly, G. A. (1955). *The psychology of personal constructs* (Vol. 1). New York: Norton.

Meichenbaum, D. (1977). *Cognitive behavior modification: An integrative approach.* New York: Plenum.

Okun, B. (1992). *Effective helping: Interviewing and counseling techniques* (4th ed.). Pacific Grove, CA: Brooks/Cole.

Patterson, L. E., & Eisenberg, S. (1983). *The counseling process* (3rd ed.). Boston: Houghton Mifflin.

Rogers, C. R. (1942). *Counseling and psychotherapy.* Boston: Houghton Mifflin.

Rogers, C. R. (1957). The necessary and sufficient conditions of therapeutic personality change. *Journal of Consulting Psychology, 21,* 95–103.

Rogers, C. R. (1961). *On becoming a person.* Boston: Houghton Mifflin.

Rogers, C. R. (1965). Client-centered therapy. In E. Shostrom (Ed.), *Three approaches to psychotherapy* (film). Santa Ana, CA: Psychological Films.

CHAPTER 4

IN-DEPTH EXPLORATION

OVERVIEW OF THE EXPLORATION PROCESS

As the counseling process continues, the counselor—at first a stranger—is increasingly seen by the client as a caring and genuine person who is able to understand and accept the client's problems and emotions. When the client begins to understand that the counselor will not reject him or her for what he or she says, the client begins to feel that it is safe to share more private thoughts. The counselor, having gained increasing insight into the client's personality and circumstances during the first stage, has a base of information that provides context for the client's continued disclosures. The counselor is therefore able to respond more accurately and completely to what the client says. Furthermore, as the level of trust builds, the counselor can begin to provide to the client feedback about his or her thoughts and actions that would have raised defenses and perhaps caused the client to terminate counseling if introduced earlier. The combination of greater client readiness and increased counselor knowledge of the client leads to the second stage of counseling—in-depth exploration—which is characterized by more emotional intensity and more complex thought processes than are present in the first stage.

As the client continues to disclose his or her intimate thoughts and feelings and describes not only his or her own behaviors but also the reactions of others, both client and counselor become more aware of the following:

- Significant events that have shaped the client's personality and present circumstances
- Conditions such as ethnicity, culture, gender, or socioeconomic status that have influenced the client's opportunities
- Deficiencies in the client's ability to cope with life circumstances

- Strengths that the client has but may not be applying to the resolution of his or her problem(s)
- Interpersonal relationships with significant others in the client's life that affect the client's thoughts, feelings, and actions (including unfinished business from the past)
- Feelings about self and others
- Goals that have been implicit in the client's unsuccessful efforts to resolve his or her problem(s) and that now can be made explicit in preparation for action planning

These new awarenesses create insights into the client's concerns and open up options for resolution of those concerns. The counselor will frequently develop such insights more quickly than the client because of the counselor's familiarity with personality dynamics and the patterns of behavior that people typically employ to fulfill their needs when confronted with circumstances that block the way. These counselor insights form the basis for diagnosis, discussed in more detail in Chapter 5.

It is the counselor's diagnostic thoughts that create the basis for the additive responses of second-stage work. Such additive responding makes connections among things the client has disclosed or that are implicit in the client's statements. These connections have not been made explicit by the client, who may not be aware of them. (Chapter 6 includes a discussion of the counselor's responsibility to make continuous clinical judgments about the client's readiness for additive responses.)

These are the crucial goals of the second (in-depth exploration) stage of counseling for the client:

- To gain insight into his or her strengths, deficiencies, interpersonal functioning, "baggage" from the past, feelings, desires, and needs
- To begin formulating goals regarding changes that he or she has the power to make and that will lead to more complete satisfaction of desires and needs

The belief that insight can serve as a basis for client growth and problem resolution is fundamental to this book's generic model of counseling.

With children and with lower-functioning adolescents and adults, the gaining of insight may sometimes be an unachievable goal. Looking inside is not a skill that all clients possess. Even so, helping clients to become more internal in their orientation—to consider the thinking behind their behaviors—is a worthy endeavor.

It is true that some clients may experience emotional release (catharsis) through counseling that leads to more satisfactory feelings and behaviors *without* ever achieving a thorough understanding of the roots of their difficulties. Likewise, some clients achieve insight into the nature of their difficulties without ever formulating goals or taking action. Those clients may not experience much relief of their distress or resolution of their problems as a result of the counseling experience.

Not all clients need the same kind of insight. Some may need a clearer picture of interests and ability patterns that relate to career choice. Some may need clarification about how their behavior affects others. Confused clients may need help understanding internal beliefs and perceptions that create conflict. Still others may benefit from becoming aware of their inner emotions and how these emotions are expressed, disguised, or stifled.

Insight (or interiority) is gained through the process of exploring significant themes, patterns, concerns, and issues. During the first stage, some initial nuclei are identified. In the second stage, they become the focus of attention and are explored in depth. As the client explores and the counselor provides feedback, elements of the client's thoughts that have been in the background emerge into the perceptual foreground. During this process, assumptions, beliefs, emotions, motivations, and inconsistencies become clearer to the client. The experience can be enlightening and tension-producing at the same time. It is important for the counselor to continue to use the core skills of the first stage to create and maintain closeness and trust while at the same time interspersing the more challenging skills of the second stage.

As Table 2.1 indicated, counselor skills and techniques associated with the process of in-depth exploration include advanced empathy, immediacy, constructive confrontation, interpretation, and role playing. The common characteristic of each of these techniques is that they provide potentially beneficial feedback to the client. To understand these skills and their relationship to the counseling experience, one first needs to understand the experience of receiving and giving feedback. "Feedback is concerned with providing clients with clear data on their performance and/or how others view them" (Ivey, 1988, p. 256).

Exercise

Everyone receives feedback throughout his or her life. Some of it is encouraging and rewarding—as, for example, when a significant other makes a statement such as "I like how you look" or "I appreciated your observations at our staff meeting." Other feedback takes the form of criticism—for example, "You would have been more effective if you had listened instead of changing the topic."

List on a piece of paper three recent experiences in which you have received feedback you regarded as critical, leaving space after each entry to write the answers to these three questions:

1. What did you notice about the giver's tone of voice and attitude toward you?
2. What were you aware of inside yourself while you were receiving the feedback?
3. What, if anything, did you do as a result of the feedback?

Repeat the same steps recalling instances when you received positive feedback. Then, before proceeding to the next section, generate from your experiences three to five principles that describe how you think people respond to feedback. If possible, share your thoughts with other students in a small-group discussion.

Principles Governing the Use of Feedback

Feedback is hard to receive. All feedback, even when supportive, can meet with some resistance. When receiving supportive feedback, people often have mixed emotions: they appreciate the good words yet feel uncomfortable in the spotlight of attention. Critical feedback identifies behavior that an observer has found to be inappropriate, and it is even harder to accept. When others are critical, one feels that one's judgment is being questioned—and this is especially uncomfortable if one is already feeling uncertain or guilty about some action.

Feedback that does not fit a person's self-image will be harder to receive than feedback that is consistent with that self-image. If a client tells a counselor that he or she appreciates the counselor's insights and sensitivity, the counselor can accept the feedback. In contrast, if a client tells a counselor that the counselor has not understood his or her concerns, the counselor's reaction is apt to be defensive: he or she may try to prove the client wrong by restating the hypotheses about the client's motivations and behaviors. Supportive feedback that does not fit a person's self-perceptions is also hard to receive, though the affective experience is different. A counselor may feel anger or disappointment at having his or her skill underestimated but may feel embarrassed by an overestimate.

Feedback takes time to internalize. Feedback is never fully internalized at the time it is received. Elements that are consistent with the person's self-concept are quickly acknowledged, other elements are rejected or discounted, and still others are reviewed many times before they are recognized as valid and incorporated into a revised self-concept. Even when a person wants feedback, he or she may resist it (Harrison, 1973).

Feedback is easier to receive if it comes from a trusted source. As discussed in Chapter 3, trust develops in the context of a genuine, caring relationship in which a counselor shows respect for a client. Trust is also a function of the client's assessment of the counselor's credibility. If the client sees most of the counselor's observations, perceptions, and judgments as valid, he or she will work harder to accept new feedback that may initially seem incongruent with self.

Feedback is easier to receive when the giver offers it with a calm presence. The calm presence of the counselor supports the client in assimilation. A counselor who is anxious, angry, or embarrassed while giving feedback will cause the client to feel insecure and will block assimilation. Counselors may have special difficulty with giving feedback when they project themselves into the role of the receiver and in essence say to themselves, "I would not want to receive what I am giving."

Feedback is more effective when it is communicated clearly and specifically. To offer feedback clearly, a counselor must know what information he or she wants to give and be very comfortable about giving it. When the counselor has not fully formulated the information, it is apparent that the client will not be able to perceive what is not stated. If the counselor knows what he or she wishes to say but feels uncomfortable about saying it, the language will often be obscured in an effort to take the edge off the message. The counselor's defensive communication can distort the message, and the client is deprived of the counselor's accurate insight. Feedback that is specific will promote more effective exploration than feedback that is vague (Ivey, 1988). It is much more helpful to say "You feel that your husband does not respect your career and this angers you" than it is to say "You feel that your husband keeps putting you down." The first response encourages discussion of specific hurtful exchanges between husband and wife that can be brought into the here and now; the second response does little to focus the discussion.

Feedback can only be absorbed in small doses. Although most people have many areas that would profit from change, they cannot work on all of them at one time (Ivey, 1988). It is important not to overwhelm the client. The urge to present the client with a litany of shortcomings is greatest when the counselor is feeling frustration with the progress of the counseling.

Feedback is presented for client consideration, not as indisputable truth. Consider, for example, the case of a boy who repeatedly gets involved in fights with other boys on the playground. These incidents occur when other boys call him names or make remarks about his mother. The other children have learned that he responds excessively to their taunting and have made a game of antagonizing him. This is an example of a feedback statement that is quite interpretive: "You defend yourself and your mother at the drop of a hat. You seem to feel there is something wrong with your family." Although it might generate insight for the boy to consider this hypothesis, he may also have trouble acknowledging it if he is sensitive about having a single parent with a live-in boyfriend. Even if the hypothesis is correct, it will not help if the client is not ready to accept it, and nothing will be gained by forcing the issue.

Lynn and Frauman (1985) state that behavioral feedback, based on observation, is generally easier for clients to accept than inferential feedback. In the above example, it would probably be much easier for the client to accept that he responds with a short fuse to the taunting of other children (observation) than it would be for him to face what it is about his family that he feels he must defend (inference).

Since feedback is presented for client consideration, the counselor follows the giving of feedback with an assessment of whether the feedback was received and whether it proved useful to the client. Remember that, as we stated above, even though feedback may not appear to be received at the time it is given, it may in fact be reviewed by the client and brought up again later, after he or she has become used to the idea.

Importance of Feedback in the In-depth Exploration Stage

Feedback helps people grow and learn about themselves and their environment. As clients learn more about their present behaviors, they become clearer about their goals for change. But people tend to be in conflict about feedback; they want it for growth and resist it at the same time.

The essential counselor work of the in-depth exploration stage is providing clients with feedback about themselves in such a way that it can be assimilated and used for growth. Since feedback will be resisted, offering it is an art form. Working in line with the principles just developed will increase the chances of assimilation and use. The next sections discuss several modes of providing feedback: advanced empathy, immediacy, confrontation, interpretation, and role playing.

ADVANCED EMPATHY

In Chapter 3, empathy was described as the "counselor's ability to enter the client's phenomenal world—to experience the client's world as if it were [his or her] own without ever losing the 'as if' quality" (Rogers, 1961, p. 284). Primary empathy, the level appropriate for the initial-disclosure stage, was described as responses indicating that the counselor has understood those themes that are readily apparent. Nonverbal communication and interchangeable responses were discussed as important modes of communicating primary empathy. The in-depth exploration stage is characterized by a deeper level of openness in the client—an increased readiness to explore significant themes and to become aware of the less obvious meanings behind those themes. Concomitantly, the counselor's level of empathy must change from primary to advanced.

Advanced empathy includes the sensing by the counselor of what the client has implied but perhaps not directly stated in his or her disclosures. Truax (1967) indicated that at an advanced level of empathy, the counselor "uncovers the most deeply shrouded of the client's feeling areas of which he [or she] is scarcely aware. [The counselor] moves into feelings and experiences that are only hinted at by the client" (p. 566).

Martin (1989) uses the term *evocative empathy* to indicate that at any stage of counseling, the counselor should give voice to the client's "intended message." However, he also indicates that "later in therapy the therapist can go a lot further and say much more emotionally poignant things—because the client intends the therapist to hear so much more" (p. 6).

Advanced empathy is not to be confused with interpretation; empathy is always based on the client's frame of reference (Egan, 1990; Martin, 1989). The empathic therapist sees the client as the source of meaning and feeling. When conveying advanced empathy, the counselor uses statements whose purpose is to evoke the feeling and meaning that already reside in the client, even though the client may be barely aware of their existence. With interpretation, the assumption is that the counselor has an insight that is more perceptive than the client's and that should be shared.

Egan (1990) poses four questions that counselors can ask themselves to identify material for the additive responses that characterize advanced empathy:

- What is this person only half saying?
- What is this person hinting at?
- What is this person saying in a confused way?
- What messages do I hear behind the explicit messages? (p. 214)

An example wil help illustrate the difference between primary and advanced empathy and show the differential impact of the two kinds of response. The client in this example is a middle-aged, white female in an in-patient psychiatric unit being treated for depression accompanied by suicidal ideation.

CLIENT: My past caught up to me and really floored me. I've been trying to keep my past *in* the past.

COUNSELOR 1: It's hard to talk about the past.

COUNSELOR 2: Keeping the past bottled up hasn't worked. As difficult as it is to talk about it, you know you have to do it.

The first counselor used a primary-empathy response—restatement of the idea that it is difficult to talk about the past. Some purposes are served by this response. The client learns that the counselor understands that the process is emotionally sensitive. The client is also prompted to think again about how difficult it might be to talk and can possibly sense enough caring in the counselor to go on. The client feels some understanding from the counselor but has a lot of choices about how to continue—which is appropriate early in the counseling process.

The second counselor used an advanced-empathy response. While recognizing the client's caution, the counselor sensed her intended message—that it is time to get some of the past out on the table, however difficult it may be. This interaction took place a couple of days after admission to the unit and after considerable therapy had already taken place. Following the advanced-empathy response, the client proceeded to describe that she had been sexually abused many years ago by her brother. The incident that precipitated the current suicidal depression and hospitalization was the client's learning that this same brother had recently molested her 16-year-old daughter. The client was suffering from guilt that she had never confronted the earlier attack and had thereby set the stage for her daughter's traumatic experience.

Note that the surface message in the client's statement is that talking about the problem is difficult and, by implication, that she would prefer to let the past stay in the past. The real intended message at this stage of counseling, however, is that "it is important to me to deal with what I have been avoiding for so long."

Case: Colleen

Colleen was an adolescent whose parents were separated and who had impaired vision. She lived with her mother and occasionally visited her father at his apartment on weekends. She worked regularly with her counselor on a variety of themes. In one session, she appeared unusually depressed. The counselor observed her appearance of depression, which stimulated a discussion of her visit to her father on the previous weekend. The counselor first contacted emotions in her initial advanced-empathy response: "Sounds like your visit with your father was very stressful for you." Colleen affirmed this observation and went on to explain that her father seemed withdrawn, distant, and uncommunicative. The counselor's second advanced-empathy response addressed the client's needs: "You need your father to communicate with you, and when he does not, that makes you wonder if he really wants you to be with him." Colleen nodded her head affirmatively with signs of pain around her eyes and mouth. The counselor's next advanced-empathy response identified an important need that Colleen had been holding inside for a long time: "I sense it's really important for you to know where you stand with your father." This brought agreement as well as tears.

When sufficient exploration had taken place, the counselor summarized: "What you want is to know how your father feels toward you. Let's see if we can figure out a way for you to find out." This introduced third-stage work into the counseling process. (The third stage will be discussed in Chapter 7.) Some role playing, including role reversal, was used to help Colleen learn how to ask.

Colleen left the counselor's office appearing very unsure, but she seemed much happier the next week. She explained that she had visited her father and told him that she found it hard to know what to say to him when she visited. Her statement led to an intimate conversation in which her father indicated that he felt the same way. At one point, he said that it was hard for him to know how to be a good father. Colleen reported that this made her feel relieved because she knew what to say. Her response to her father, "Let me know that I'm important to you," appeared to make an important transition in their relationship.

A key response in this episode was the clarification of Colleen's needs. This led to the establishment of a goal and strategies to fulfill that goal. She was able to identify a sense of personal power, which in later counseling was related to her sense of helplessness felt because of her visual difficulties.

Questions for Further Thought

1. What problems in coping might you expect Colleen to have as a result of her impaired vision? What implications do your assessments have for the development of advanced-empathy responses?
2. What constitutes a "healthy" adjustment to the separation of one's parents?
3. A client's tears are a frequent part of the counseling experience, especially when the process reaches the second stage. What are your feelings when another person cries in

your presence? How do your inner feelings influence how you respond? Would the gender of your client make a difference in your reactions? What should a counselor do when a client cries?

IMMEDIACY

"Immediacy refers to the current interaction of the therapist and the client in the relationship" (Patterson, 1974, p. 83). Thus, an immediacy response is a communication that provides feedback to the client about the counselor's inner experience of the relationship at a given moment. Immediacy responses provide opportunities for client and counselor to explore together any stresses in their relationship, any ways in which that relationship resembles the client's relationships outside of counseling, and the counselor's reactions to stress the client shows when discussing particularly painful material. In the simplest form, immediacy responses might provide an opportunity for the counselor to observe that the client's spontaneity in a given session seems different from previous sessions or, perhaps, that it feels like the counseling has "bogged down." A client of different ethnic or racial background or gender may show discomfort that will prompt the counselor to make the issue of client-counselor differences a topic for discussion. Any time that issues of the client-counselor relationship that are experienced by the counselor as potentially disruptive are introduced for direct discussion, this is a form of immediacy response.

Immediacy responses often take the form of "I" messages and are offered to stimulate exploration of some significant theme, not as criticism or punishment. Consider the following example of the use of immediacy responses with an adult male client. The initial statement came about twenty minutes into the session.

COUNSELOR: Tom, for the last five minutes or so, I have not been feeling comfortable about what is happening. To me it seems as though we are circling around the theme of you and your father without really getting into it.

CLIENT: (*after a pause*) I think you are right. So what.

COUNSELOR: For the last two sessions, I have felt as though we have circled without focusing. Today I decided I want to say something.

CLIENT: If that's how you have felt, why didn't you say something sooner?

COUNSELOR: I haven't been sure whether you would be open to hearing that message from me.

CLIENT: You must see me as pretty defensive.

COUNSELOR: All of us, including me, are defensive in some way. I can be excellent at intellectualizing.

CLIENT: I've noticed that sometimes. But you seem to catch yourself and not let it get too much in the way.

COUNSELOR: Would you be willing to tell me about my defenses?

CLIENT: Sure. *(The client proceeded to describe his observations of the counselor's defenses. In the midst of this description, the client started telling how the counselor's intellectualizing occasionally moved the focus away from significant material. Suddenly his face showed signs of insight.)* Am I really talking about what *I* am doing?

As Patterson (1974) explained, "Immediacy is important because the client's behavior and functioning in the therapy relationship are indicative of his (or her) functioning in other interpersonal relationships" (p. 83). If, for example, the client exhibits unusual dependency in the counseling relationship, then it is likely that she or he may do the same with significant others. Helping a client to recognize, understand, and manage dependency in the counseling relationship should generalize to the client's other relationships as well.

Sometimes, too, the counselor's experience with the relationship is directly related to the material being discussed and the client's strategies of avoidance. By providing feedback to the client about his or her behaviors, the counselor can often help the client to acknowledge discomfort and make a more conscious decision about how to proceed. Consider the following example of an adult female client.

COUNSELOR: Right now I am feeling very cautious. Part of me wants to explore how things are between you and your boyfriend, and part of me is holding back. . . . I wonder if you may be having a similar experience.

CLIENT: *(pause followed by slow speech)* Yes. This whole thing is so screwed up for me that sometimes I would just rather not think about it.

COUNSELOR: I get from that that you would feel more comfortable if we talked about something else.

CLIENT: *(still slowly)* It would be safer . . . but I still need help sorting this all out.

COUNSELOR: I have a sense of how loaded this area is for you. . . . Would it help you if I just listen while you talk? *(At this point, the client inhaled and started to tell her story.)*

Our own experience as counselor educators, along with the clinical experience of people such as Jourard (1971), leads us to believe that immediacy is one of the harder counseling skills to develop. Although learning effective words can help in developing this skill, immediacy communication is difficult because it is often threatening. Immediacy requires intimate communication in which the counselor discloses part of himself or herself to the client. Trainees may fear such self-disclosure because of a lack of experience, instances of being hurt in previous intimate encounters, or concern about hurting the client. Some have trouble attending to their own inner experiences and thus cannot share such experiences with others. Remember that the purpose of counselor self-disclosure in immediacy responses is to assist the client in his or her self-exploration, not to allow the counselor to ventilate pent-up emotions. Counselor self-disclosure can also shift the focus of the discussion from the client to the counselor if it is not skillfully accomplished.

To work on your immediacy skills, you will need to identify your own possible sources of fear. The assistance of a supervisor or peers in group supervision can promote growth in the ability to use immediacy skills by providing feedback about passed-over opportunities for immediacy responses.

CONFRONTATION

Confrontation is one of the more controversial counseling techniques partly because the everyday use of the word often implies conflict or attack—for example, "Pro-life and pro-choice demonstrators confronted each other on the Capitol steps today." Implied are dimensions of anger, violence, struggles over power and control, and an adversarial rather than a cooperative relationship.

Confrontation as used in counseling is defined as a counseling intervention that describes a client's discrepancies, contradictions, and omissions. Confrontation is done *for* and *with* the client not *to* and *against* the client. In other words, the counselor does not confront to satisfy his or her own needs, to vent feelings of frustration with the counseling, or to punish the client. Instead, the counselor holds a sincere belief that the client will experience growth by paying attention to some discrepancy or incongruence that he or she has revealed. Egan (1990) proposed that the word *challenges* may more correctly describe the intent of responses that are commonly called confrontations.

The first step in the confrontation process is the identification of material in the client's statements and behaviors that requires further attention through observation, questioning, and listening (Ivey, 1988). "These observations are then fed back to the client in a clear, concise, nonjudgmental manner" (Ivey, 1991, p. 93). The subsections that follow first describe some of the possible discrepancies, contradictions, or mixed messages that may be confronted (Dinkmeyer & Dinkmeyer, 1985; Egan, 1990; George & Cristiani, 1990; Ivey, 1988, 1990) and then consider the special sensitivities required to confront client defenses.

Discrepancies between Client Perceptions and Accurate Information

People act on the basis of what they believe to be true, and when their beliefs are inaccurate, they act in ineffective ways. Helping a client correct misconceptions can lead to more rewarding behavior. For example, an adolescent female may express the belief that washing with soap and water after intercourse eliminates the possibility of pregnancy. She is acting with serious misinformation; helping her understand reality more accurately may help her take more appropriate preventive actions. Her counselor may choose to confront her as follows: "Rina, it sounds as though you believe washing with soap and water will make sure you don't get pregnant. (Rina nods affirmatively with an anxious look.) There are several ways to prevent pregnancy if you want to, but washing with soap and water is not one of them. Would you find it helpful for us to discuss the possible ways?"

Discrepancies between Client Expectations and Likely Possibilities

Counselors who work from a rational orientation help clients think rationally and sensibly about what is happening to them. A counselor who listens carefully will often hear statements that reflect irrational thinking—for example, "I must be accepted and loved by everyone, or I cannot be happy." Such thinking is said to be irrational because common experience shows that no one is loved by everyone, and yet many people are able to achieve happiness. (See the section on cognitive counseling in Chapter 12 for further discussion of this kind of confrontation.)

Discrepancies between Verbal and Body Messages

Perls (1969) indicated that although it is relatively easy for people to use words to cover up internal truths, body messages provide more accurate information about what is happening within. Providing a client with feedback about body messages or about discrepancies between verbal statements and body messages can lead to more honest communication. For example, a female client was speaking in a slow, labored, drawn-out fashion. The counselor asked the woman if talking about the particular theme was all right for her. While saying yes, the client crossed her arms and pressed them to her stomach. The counselor called her attention to her closed posture: "You say that you are comfortable with this but your arms are crossed in close to your body, and you look tense." This led to some important exploration of the reasons why the theme was hard for her to talk about.

Sometimes, too, clients try to hide their feelings about what they are saying by effecting a contradictory physical message. Perhaps the most frequent example of this is smiling while discussing or describing some very distressing event. The true emotion may be fear, anger, or embarrassment, but the smile is presumed by the client to make the message more acceptable. A simple observation such as "Are you aware that you are smiling as you talk about this unpleasant topic?" will often elicit more honest and therefore more productive communication.

Discrepancies between Behaviors and Stated Goals

Perhaps one of the most usual manifestations of a discrepancy between behaviors and goals is seen in persons who seek to be accepted by others but whose behaviors make such acceptance very unlikely. On the elementary school playground, this pattern is observed in the child who seeks positive attention by tormenting his or her peers but ends up without any friends. The case of Lionel presented later in the chapter also illustrates such a pattern of self-defeating behavior.

Contradiction between Statements and Actions

Consider the case of an adult female client who was trying to cope with a separation initiated by her husband. She described herself as a good wife and expressed intense agitation and confusion about why her husband wanted a divorce. In a later session, she told of an affair she had shortly after her marriage. By presenting two pieces of information together in the same sentence, the counselor provided the client with feedback about the incongruence of her self-perception and her actions as seen by her husband: "You see yourself as a good wife, and yet you also had an affair shortly after you were married." This led to exploration about her concept of being a good wife, her standards of sexuality, and what the affair had meant to her husband, who had found out about it.

Mixed Messages

Explore your own experiences for mixed messages that you have received from others. One of the common kinds of mixed message is conditional love: "I love you when you do what I want." Such messages are commonly delivered to children by parents and between spouses. The sender of the message is usually not aware that demands are being communicated along with love and becomes confused that the receiver of the message does not feel loved. Confrontation with the demanding or negative part of the message creates fertile ground for further counseling work.

Working with Client Defenses

Directly confronting a client's defenses is probably the most risky kind of confrontation, since such a frontal attack has a high probability of raising the client's resistance even further. Consider the adolescent client with controlling parents who resists parental control by behaving in a surly and uncooperative manner. This defensive behavior serves the purpose of demonstrating that the parents cannot control the client's behavior, but it has the side effect of keeping family life in a state of turmoil that is usually as aversive to the youngster as it is to the parents. In order for a more satisfactory relationship to be developed, the adolescent should be helped to understand the defense, and direct confrontation may not accomplish that end.

It is important to recognize that a client's defenses may be longtime friends; often they have helped the client cope with stresses and pains of life over a long period (Clark, 1991). However, a person's defenses may not only serve him or her well but may also result in forms of self-defeat and misery. Even so, the client will often resist giving up defenses. The task of the counselor is to *work with* the client's resistances rather than attacking them (Polster & Polster, 1973). Rather than demand that resistances be given up, the counselor must simply help the client see and experience them with greater clarity—become more familiar with his or her longtime friends. In the example above, the counselor might tentatively test the client's ability to see his or her surly, uncooperative behavior as a defense: "Could it be that you act so angrily so that you won't feel as controlled?"

Working with defenses or resistances is only confrontational in the sense that the counselor's interventions are designed to provide useful feedback to the client. The effectiveness of the process is reduced and may even be counterproductive if any sense of attack is involved.

Since resistance is often accompanied by physical responses of tight muscles, constricted breathing, rigid posture, failed eye contact, and other signs of discomfort, drawing the client's attention to such a physical manifestation can be helpful. This can be done by describing the activity, asking the client what the resisting muscles would say if they could talk, slightly exaggerating the behavior, or using one's own body to show the client what he or she is doing. To be effective with these interventions, the counselor must learn from experience with each client which kinds of approaches are most acceptable.

Guidelines for Constructive Confrontation

A serious problem of the confrontational process is that the client may become confused about whether the counselor is an ally or an adversary.

When a counselor confronts a client, even with the noblest of motives, the client is not sure of the counselor's intent. He or she may be an ally, in the sense of wanting to encourage the client, but may come across as an enemy of the client's resistances. It is helpful if the counselor keeps the focus on the fact that all confrontation is *for* and *with* the client not *to* or *against* the client.

We suggest the following guidelines for confrontations:

1. Remember that confrontation is not the only mode of giving feedback to a client. Use it sparingly.
2. Never be an enemy to your client.
3. If you are feeling angry toward your client, that's your problem. You won't help by punishing. If you are angry, you are probably reacting to one of your client's resistances. You can use your anger as a source of information to recognize your client's resistance. If you have trouble resolving your anger, seek supervision or consultation with a trusted colleague.
4. Be very clear about your reasons for confronting. If you have a plan for what you are trying to accomplish with the client, it is much more likely that confrontation will be based on the client's needs and not your own.
5. Be a total ally of the client, not an ally and an adversary. This means that confrontation should be a means of providing feedback and that it should not contain staged or implied disapproval.
6. Use direct and simple language. Vague language may mean that you are unclear about what client material you want to confront, your motives for confronting, or whether the client is ready to hear what you want to say. Under any of these conditions, the time is not right for confrontation.

Case: Lionel

Lionel was a college student seen by a male counselor in a college counseling center. His initial concern was a feeling of depression about being alienated from the other students on the campus. He reported having a few acquaintances, but no friends, male or female. The pain of his circumstances provided high motivation for work.

During the initial-disclosure stage, Lionel and his counselor discussed how it was for him to be in college and what he noticed happening when he did talk with fellow students. During these conversations, the counselor noticed a pattern of speaking disparagingly of others. Lionel seemed to be quite cynical.

When this pattern continued, the counselor decided to confront. His confrontation was the critical moment that moved counseling into the second stage. "Lionel, I have often heard you make cynical statements about people in your classes. You say that you want to make friends, but you say things that are not friendly." This led to some intense exploration of how Lionel's cynicism might affect others with whom he wanted to develop relationships. It also opened up a pattern about which he previously had only a dim perception.

During the third stage, the counselor helped Lionel work on making more positive and supportive statements about others. Making this transition turned out to be difficult. The counselor concluded that Lionel was maintaining his cynicism because he was getting something from keeping it. Counseling moved back into the second stage, and further exploration disclosed that Lionel used his cynicism to maintain a relationship with his father. Lionel felt that letting go of his cynicism could result in alienation between himself and his father, a fear he had unknowingly been holding onto for a long time. Role playing helped Lionel develop important awarenesses of what was happening in his relationship with his father, what he wanted in the relationship that he was not getting, and what he feared losing. These awarenesses, though painful, helped Lionel conclude that the pattern of cynicism that he used to protect himself in his relationship with his father was not helping him in his peer relationships.

Moving again to the third stage, Lionel and his counselor did some training in direct interpersonal communication, in which Lionel's task was to make honest and positive statements to the counselor. The counselor gave supportive feedback to Lionel's effective communications. Putting the principles into practice in his classes and in social situations, Lionel reported that he had made some friends and had two dates that went fairly smoothly. Counseling terminated when Lionel reported that he had been feeling much more comfortable in his peer relationships.

Questions for Further Thought

1. The second stage involves helping clients see something about themselves more clearly. What specific things did Lionel see more clearly as a result of his counseling experience?
2. How do you think Lionel felt when he received the feedback from his counselor? How do you think the counselor felt about offering it?
3. How do you react to cynical people?

INTERPRETATION

Responses that promote in-depth exploration all serve to help the client add perspective to his or her conception of self in the environment. Advanced empathy was considered to be additive in the sense that the counselor included in his or her response some material of which the client was only vaguely aware and had only *implied* in previous statements. Confrontation is additive in the sense that the counselor pulls together elements of the client's story and in-session behaviors that are incongruent, in order to enable the client to integrate the inconsistencies. Interpretation is also a form of additive responding, the purpose of which is to "add meaning to the client's attitudes, feelings, and behavior" (George & Cristiani, 1990).

Interpretation is often barely distinguishable from advanced-empathy responses (Martin, 1989; Patterson, 1974). This is true because the counselor's hearing is attuned by his or her theoretical knowledge in such a way that what he or she hears and responds to will be consistent with either type of statement. Martin (1989) finds the difference between advanced empathy and interpretation in whether the response stays with the client's frame of reference or introduces a new frame of reference. According to Ivey (1988), in using interpretation, "the interviewer presents the client with a new frame of reference through which to view the problem or concern and, hopefully, better understand and deal with the situation" (p. 259). Some therapists prefer to use the term *reframing* instead of interpretation, partly because interpretation has gained a reputation of being a mystical technique used by "depth" therapists.

Although we present interpretation as one of the techniques that can help clients discover deeper meanings, no mystery exists about the basis for the additive component of the counselor's response. Interpretation involves placing meaning on observational data using *theory* (Brammer, Shostrum, & Abrego, 1989). When interpreting, the counselor presents the client with hypotheses about relationships, meanings, or behaviors that emerge from the counselor's *theoretical* understandings of human personality.

The frequency with which a counselor uses interpretation and the nature of the interpretations offered therefore depend upon the theory or theories to which the counselor adheres. Person-centered counselors regard the client's frame of reference as the only one that matters and generally see interpretive responses as errors. In the psychoanalytic approach, interpretation is seen as the process through which clients gain new insight into their symptoms and dynamics, and it is thus seen as necessary to the change process (Garske & Moltena, 1985). Psychoanalytic counselors use a system based on the work of Freud (1954) for understanding clients and generating hypotheses. On the other hand, a counselor who sees client meanings, feelings, and behaviors as a function of thought processes will generate hypotheses based on observations about the client's faulty thinking.

Consider the example of a 24-year-old male who presents for counseling with a history of having lost a half-dozen jobs since graduating from college three years ago. In each instance, the client describes having been supervised very closely, treated like a child, and evaluated unfairly. Each time, the client either quit or was fired after repeated squabbles with his supervisor. From the client's frame of reference, the problems with each of the positions were due to bad luck in having such an ineffective supervisor. Although that is possible, most counselors would consider the odds of finding a half-dozen such poor situations unlikely and would begin looking for what the client may have contributed to the disputes that ensued. This is one possible additive response: "Have you thought about ways in which your relationships with your supervisors in all these situations seem similar?" Such a response is mildly confrontative but, by and large, stays with the client's frame of reference. However, if the counselor has also learned that the client had a stormy relationship with his father while growing up and continues to harbor anger toward his father, this might be a possible response: "Do you see how similarly you seem to feel about your employers and your father?" This response is based on the hypothesis that the client may be projecting elements of his relationship with a strict and controlling father onto his supervisors.

Even counselors who believe that interpretation is a useful technique agree that it does not work if used very early in the counseling relationship. Precisely because interpretation asks clients to think very differently about the causes and sources of bad experiences and bad feelings, groundwork in both relationship and content must be laid before it is likely to work. In effect, the client is asked to consider substituting the counselor's interpretation about life events for his or her own previously held view. Premature interpretation has the effect of casting the counselor as insensitive and of prompting the client to hang on more tightly to those defenses that have helped him or her construe reality in safe but ineffective ways. Just as interpretation is not used frequently in the early stages of counseling, it is used less frequently as a single counseling session or a series of sessions moves toward closure, so as to avoid opening issues that cannot be adequately integrated in the remaining time.

ROLE PLAYING

Role playing can be especially useful in solving difficulties with interpersonal relationships and has application in both the second and third stages. In the second stage, the focus is on increasing awareness of self and others and setting goals; in the third stage, action planning is the main agenda. In practice, the two stages often run together when role playing is used.

The principal value of role playing is that it can bring into the present events that have happened in the past or events that are anticipated by the client as possible in the future. Role playing can be set up in various ways. Often the counselor plays one party to a conflict, perhaps a sibling, spouse, or parent of the client, while the client plays himself or herself. The counselor can also play the client, freeing the client to play the other party. Or, with use of an empty chair, the client may take on both roles—himself or herself and the significant other—changing chairs between responses and delivering both persons' lines.

Setting up the role play is an important part of the activity. After carefully explaining how role playing might help, the counselor will find it useful to give the client the role of drama director. That is, the client decides (with help) who plays which role, instructs the counselor on the characteristics of the person he or she is playing, and decides how to play his or her own role. Details about the setting should be clarified by the client to achieve as much authenticity as possible. It may be useful to have several "rehearsals" to get all the details right and to portray personalities accurately. The rehearsals themselves deepen the client's understandings of the parties involved, while at the same time informing the counselor.

During the role playing, the counselor has several tasks: to play his or her own role accurately, to pay close attention to the client's behavior and emotions, and to pay close attention to his or her own inner experiencing. When the role playing is completed, the client and the counselor discuss what has occurred. The counselor has a rich supply of information from the experience to offer as feedback to the client. The information can come from within self and be offered as immediacy communication ("As I played your wife, I wanted to hear your ideas but was feeling talked down to and was hurt that you did not see me as an equal to you") or from observations of the client ("As I asserted my ideas, your muscles seemed to freeze"). The feedback helps the client clarify emotions, desires, beliefs about the other person, beliefs about self, and the impact of his or her behaviors. These new insights serve as the basis for developing action plans to improve the relationship in the third stage.

Case: Jack and Eve

Jack and Eve were a married couple in their mid-thirties who had been married eight years. Their only child, Tommy, was 6 years old. They came for counseling because, as Eve described it, "Our family is not a healthy place to live." Their relationship was characterized by disagreements over money, role assignments, and parenting practices. Jack felt stress at work and was short-tempered at home. Eve was primarily responsible for child care, managed the household affairs, and was not employed outside the home.

In one session, a recent episode in which Jack had yelled at Tommy was the focus of counseling. A role-playing session was established in which the counselor played the role of Tommy. The situation was as follows: Jack came home from work to find Tommy's room a mess. He hollered loudly for Tommy to come to his room immediately. Slowly and apprehensively, Tommy approached his room. Once Tommy was there, Jack proceeded to "read him the riot act." In the midst of Jack's diatribe, "Tommy" (the counselor) suddenly shut his eyes, held his hands over his ears, turned around, ran away, and hid under a piece of furniture in the next room. This infuriated Jack, who demanded in a loud voice that "Tommy" come back. In a quaking voice, "Tommy" said, "No, I'm afraid you might kill me."

The role playing ended; Jack was visibly shaken and waved his hand to stop the action. He sat down and held his head in his hands. He repeated several times, "I just never realized how badly my anger hurt." After a while, the counselor asked Eve about her reactions to Jack's anger. Her statement, "Just like Tommy's; it terrifies me, too," validated the feedback.

Discussion

Until this experience, Jack had not taken seriously the effects of his anger on others, rationalizing it away, disowning it, and arguing that other family members had to learn to accept his anger. The role-playing situation in which the counselor deliberately did what he thought Tommy wanted to do but was afraid to do helped Jack see and experience the effects of his anger. In a later session, there was a follow-up exploration of what Jack would lose by controlling his anger. This led to discoveries related to fear of losing power, fear of being the underdog, and the use of anger to intimidate. Over time, these insights seemed to help Jack reduce his expressed anger, show more patience, and express affection more easily.

Questions for Further Thought

1. What principles of role playing did the counselor use with Jack and Eve? Were there other things he could have done?
2. From the information presented, how would you describe Jack? Do you see additional counseling work that may be needed?
3. What principles about giving feedback did the counselor use?
4. Have you ever received feedback about how you are seen by another that surprised you? How did you react when you received it? Did you change as a result?

SUMMARY

The second stage of the counseling experience is a time for in-depth exploration of themes and issues related to the client's concerns. As exploration occurs, the counselor's task becomes that of helping the client develop new awarenesses and perspectives that can lead to growth, more effective coping, and clarification of goals. Feedback is the primary vehicle through which awarenesses occur. Communication modes such as advanced empathy, immediacy, confrontation, interpretation, and role playing stimulate this process. Clients will be much more likely to internalize feedback and use it to plan change if they feel safe and respected in the counseling relationship and if the counselor exerts every effort to continuously reinforce the message that he or she acts out of caring.

REFERENCES

Brammer, L. M., Shostrom, E. L., & Abrego, P. J. (1989). *Therapeutic psychology* (5th ed.). Englewood Cliffs, NJ: Prentice-Hall.

Clark, A. J. (1991). The identification and modification of defense mechanisms in counseling. *Journal of Counseling and Development, 69,* 231–236.

Dinkmeyer, D., & Dinkmeyer, D., Jr. (1985). Adlerian psychotherapy and counseling. In S. J. Lynn & J. P. Garske (Eds.), *Contemporary psychotherapies.* Columbus, OH: Merrill.

Egan, G. E. (1990). *The skilled helper: A systematic approach to effective helping* (4th ed.). Pacific Grove, CA: Brooks/Cole.

Freud, S. (1954). *Psychopathology of everyday life.* London: Ernst Benn.

Garske, J. P., & Moltena, A. L. (1985). Brief psychodynamic psychotherapy: An integrative approach. In S. J. Lynn & J. P. Garske (Eds.), *Contemporary psychotherapies.* Columbus, OH: Merrill.

George, R. L., & Cristiani, T. S. (1990). *Counseling theory and practice.* Englewood Cliffs, NJ: Prentice-Hall.

Harrison, R. (1973). Defenses and the need to know. In R. Golembiewski & A. Blumberg (Eds.), *Sensitivity training and the laboratory approach* (2nd ed.). Itasca, IL: Peacock.

Ivey, A. E. (1988). *Intentional interviewing and counseling.* Pacific Grove, CA: Brooks/Cole.

Ivey, A. E. (1991). *Developmental strategies for helpers.* Pacific Grove, CA: Brooks/Cole.

Jourard, S. (1971). *The transparent self.* New York: Van Nostrand.

Lynn, S. J., & Frauman, D. (1985). Group psychotherapy. In S. J. Lynn & J. P. Garske (Eds.), *Contemporary psychotherapies.* Columbus, OH: Merrill.

Martin, D. G. (1989). *Counseling and therapy skills.* Prospect Heights, IL: Waveland.

Patterson, C. H. (1974). *Relationship counseling and psychotherapy.* New York: Prentice-Hall.

Perls, F. (1969). *Gestalt therapy verbatim.* Moab, UT: Real People Press.

Polster, I., & Polster, M. (1973). *Gestalt therapy integrated: Contours of theory and practice.* New York: Brunner/Mazel.

Rogers, C. R. (1961). *On becoming a person.* Boston: Houghton Mifflin.

Truax, C. B. (1967). A scale for the rating of accurate empathy. In C. R. Rogers, E. T. Gendlin, D. J. Kiesler, & C. B. Truax (Eds.), *The therapeutic relationship and its impact: A study of psychotherapy with schizophrenics.* Madison, WI: University of Wisconsin Press.

CHAPTER 5

DIAGNOSIS IN COUNSELING

Diagnosis is a process of identifying and specifying the problem (or set of problems) the client brings to counseling and then deciding whether counseling is an appropriate intervention for resolving it. A diagnosis is determined through a joint process of information gathering and hypothesis testing, conducted until a tentative conclusion about the nature of the problem is agreed to by both counselor and client. This is a critical step in the counseling process because decisions about action plans and intervention strategies are built upon the diagnosis. If an accurate diagnosis has not been determined, interventions to solve the problem have little chance of success. Critics have contended that diagnosis can be used as a way of labeling the client as sick, losing sight of his or her individuality or reinforcing gender or racial bias (e.g., Stiver, 1986). For this reason, some counselors avoid the term *diagnosis* altogether, substituting words such as *assessment* or *appraisal of client needs.* In addition, Shaffer (1986) has argued that unless diagnosis is tied to treatment, it is a futile process. Our position is that if a clear understanding of exactly what is troubling the client is brought forth in counseling, the whole counseling process is more productive and goal-oriented. Diagnosis that is done poorly or incompletely can indeed lead to the ills cited by these critics.

This chapter presents a frame of reference for understanding diagnosis in counseling, a description of the components of diagnosis and the tools available to help counselors make accurate diagnoses, a discussion of the relationship of diagnosis to the trusting relationship between counselor and client, and, finally, a review of avenues for the counselor when counseling is not the best way to resolve the client's problem. We believe that this model helps prevent the problems with diagnosis so often cited by its critics.

A FRAME OF REFERENCE FOR UNDERSTANDING DIAGNOSIS

Diagnosis is a term borrowed from the medical lexicon, where it refers to the identification of a disease or dysfunction that is compromising a person's health. Its use in the counseling literature has become commonplace, but real parallels to the medical definition are limited. The primary parallel is that diagnosis in counseling is also a process of defining a problem that disturbs a client and then using that definition of the problem to decide on the appropriate goals and interventions. However, there are several ways in which diagnosis in counseling differs from the process in medicine. In medical diagnosis, there is usually a single identifiable cause to the disease such as the presence of cancerous cells, the fracture of a bone, or dysfunction of a kidney. In counseling, the causes of a person's sadness, anger, or poor interpersonal skills may be almost entirely outside that individual (such as a physically abusive parent) or may stem from a range of factors too numerous or too far in the past to identify specifically. Certain difficulties may also be a function of the developmental stage of the individual and thus normal rather than dysfunctional. In other words, problems in living simply don't lend themselves to a clear diagnostic classification as easily as medical problems.

The next major difference in the diagnostic process in counseling lies in the role and activity level of the counselor and the client. In medicine, the usual procedure is for the physician to ask the patient a series of questions about symptoms, conduct a physical examination, order relevant medical tests, and then make a diagnosis once all the information is available. Implied here is a highly active role for the physician and a rather passive role for the patient. The patient's primary responsibility is "fill in the blanks," to be compliant and cooperative with the medical experts. Once the diagnosis is made, the patient becomes more active, asking questions about the diagnosis and deciding whether to undertake the recommended treatment. In counseling, however, the diagnostic process is a joint enterprise in which a partnership develops. The client's input about the nature of the problem is actively sought by the counselor, and the diagnosis itself is usually a mutual decision. In other words, diagnosis in counseling is not a conclusion the counselor comes to independently. Rather, effective diagnosis requires mutual involvement in information gathering and hypothesis testing to arrive at a shared understanding of the problem. Part of this difference stems from the philosophical roots of counseling, which lie in humanistic psychology and the writings of Carl Rogers (1951). Rogers rejected the notion that problems in living were like sicknesses that needed the intervention of health professionals. Instead, he posited that psychological pain and dysfunction are best resolved by a partnership with a committed helper that allows the person needing help to come to his or her own assessment of the difficulties and to choose actions to resolve them. As is discussed in Chapter 12, humanistic psychologists like Rogers believe that humans are inherently self-actualizing, that is, motivated to become healthier and solve their problems. Our model of the counseling process amends Rogers' approach by making the counselor equally active with the client in defining the problems to be solved.

Diagnosis is a bilateral activity, but counselors and clients don't always become aware of the likely diagnosis at exactly the same point in the counseling process. Because the issues clients bring to counseling usually provoke feelings of pain and anxiety, clients sometimes block full awareness of the difficulties or their causes. When this happens, the client is said to be well defended, or to have strong defenses against the painful material. Defenses function to protect the psyche from content that appears to be overwhelming or threatening to psychological survival.

Thus, at times, the counselor may become aware of the most plausible diagnosis of the problem before the client recognizes it. In this situation, the counselor's role is to help the client feel safe enough to explore the difficulty further so that the defenses can be lowered and the full dimensions of the problem revealed to the client. The counselor ought not to "announce" his or her diagnosis to the client as soon as a plausible one becomes apparent; nor should the counselor begin interventions until the client's defenses are lowered. There are several reasons for this prohibition. First, the counselor could be in error and would then pursue an inappropriate intervention. It is important to remember that clients have the fullest knowledge of their own experience and that the process of exploring all components of diagnosis takes time and trust. Second, even if the counselor's diagnosis is accurate, when the client's defenses are not lowered, that definition may be rejected, the recommended intervention resisted, or the client may even terminate counseling. In this situation, the mistiming of the counselor's discussion of the diagnosis can be as great an error as a faulty diagnosis. Third, the presentation of the diagnosis by the counselor to the client implies that solving the problem is the counselor's rather than the client's responsibility. Thus, a unilateral diagnosis by the counselor can undermine the partnership essential to the effectiveness of the entire counseling process.

Diagnosis in counseling differs from medical diagnosis in a third important way in that it often means discussing several other significant people in the client's world. In fact, it may be that the person who makes the appointment to see a counselor is not the primary client at all. The person who makes the appointment could be a parent, friend, or partner of the one suffering the most distress. (Of course, the person in the office may still benefit from counseling; the point is that counselors cannot assume that that individual has the greatest need for counseling.) Even when the person in the office is the one in difficulty, the cause of the problem may be outside that individual. For example, a gifted Latino middle school student may be seeing a school counselor because of frustrations at school and dislike of his teacher. The process of exploring the situation may lead to the conclusion that a large part of the problem is the teacher's prejudice against persons of Mexican background. After all available evidence has been gathered, the clear assessment is that her bias, more than any other factor, has resulted in the student's dislike of the teacher and his placement in the wrong ability group. Although the student may benefit from counseling in dealing with his negative feelings and acquiring better ways of addressing his problems with this teacher and other prejudiced individuals he may encounter in his life, effective diagnosis in this case also mandates that the counselor attend to the teacher's behavior and her misplacement of this student. Perhaps the counselor may need to have the student transferred to another class. In other words, a counselor who assumes that the person in the office at the moment is the only potential client and has caused or can solve the problem is risking a diagnostic error.

Diagnosis in counseling is also somewhat dependent on the theoretical orientation of the counselor. Along with presenting a system for intervening to help clients cope better, counseling theories provide models for defining normal and abnormal behavior and for describing how and why dysfunctions develop. A counselor whose theoretical orientation rests in the behavior therapy tradition will define the problem the client presents differently than will a humanistic or cognitive counselor. The behavioral counselor will also tend to use substantially different counseling interventions than will a humanistic or cognitive counselor. As Brammer, Shostrom, and Abrego (1989) suggest, "theories act as 'templates' that indicate the client behaviors they consider most important" (p. 149). During the counseling process, counselors receive an overwhelming amount of information (both verbal and nonverbal) from the client. Theories help to organize that information and make it sensible. The process of deciding what theoretical orientation or combination of theories provides the most adequate representation of human development and dysfunction is a long and complicated one for a beginning counselor. Clarity about one's theory of counseling comes with education, supervision, and practice with real clients. Beginning counselors ought to test out different theories of counseling and approaches to diagnosis in order to judge for themselves the adequacy of each model and its appropriateness for particular clients. In time, with experience and reflection upon that experience, counselors determine the theory that best fits their own assumptions about human nature and the available evidence. Of course, given the evolving nature of theories and the research evidence regarding their veracity and effectiveness, the process of defining a theory of diagnosis and interventions must continue throughout a counselor's career.

COMPONENTS OF A GOOD DIAGNOSIS

There are six components to the information-gathering and hypothesis-testing process of diagnosis. Taken together, these components flesh out the problem(s) and give counselor and client the fullest insight into the circumstances that provoked the client to seek help.

The first component is an understanding of the boundaries of the problem, that is, the scope and limits of the difficulty the client is experiencing. Attaining clarity about the scope of the problem involves understanding its boundaries in current functioning as well as its history and duration. For example, if a client enters counseling because of feelings of loneliness and social isolation, identifying the boundaries of the problem means the exploration of exactly how socially isolated the person is and how limited his or her close connections with others are. For one person, lonely feelings can arise when a rich and extended social life has been temporarily reduced to a fairly ordinary number of social contacts because of a new living situation; for another, such feelings can be the result of an entire life of social distance and trouble with intimacy. In the first situation, the boundaries of the problem are limited, and its solution fairly straightforward. The second situation has very wide boundaries, making its solution more complicated and time-consuming.

The second component of diagnosis is the mutual understanding of the factors that maintain and lessen the problem. It is unusual for a difficulty to be experienced at a uniform level at all times. The discomfort is worse at some times, better at other times, and does not appear at all at still other times. The goal of diagnosis in this aspect is to understand whether there is a pattern to such variations and to identify jointly the factors associated with them. Coming to recognize the pattern of the problem also helps the client to view the situation as less overwhelming and more manageable. There is reassurance in the discovery that the difficulty is not randomly occurring, and the pattern makes its causation clearer. For example, a college student who feels shy and socially uncomfortable may come to notice that the feelings of shyness are worse in unstructured social situations than in structured academic or athletic situations where her role is fairly well-defined. She may also come to see a pattern of greater shyness after visits home to her critical and demanding parents or when she is with persons in whom she has a romantic interest. Taken together, this information helps client and counselor better define action plans to resolve the episodes of shyness. Through understanding the pattern of the problem, the client may also get better control of her feelings of helplessness.

A mutual understanding of the intensity of the feelings surrounding the difficulty is the third component of diagnosis. For example, when a client talks of feeling angry and frustrated with his disabled child, the goal is to identify the feelings involved and get a clear sense of the dimensions of those feelings and the associated behaviors. Does the father feel impatient and then withdraw from the child, or does he feel deep rage that makes him want to strike out and hit the child? Knowing the intensity of the feelings is important to the client's self-understanding and to the choice of appropriate counseling interventions for the problem. When clients enter counseling because of affective difficulties such as feelings of sadness, anger, anxiety, or emptiness, the exploration of the intensity of these feelings is a natural part of the first stages of counseling. However, when a client seeks problem-solving skills, assistance with career decision making, or help in stopping habitual behavior, discussion of the intensity of the feelings surrounding the issue may be neglected. However, knowledge of the intensity of feelings is also important in these cases because it can affect the commitment to change and the difficulty of relinquishing old patterns.

The fourth component of diagnosis is an understanding of the degree to which the presenting problem influences functioning in other parts of the client's life. The aim here is to ascertain how circumscribed or diffuse the difficulty is and to clarify the degree to which it is compromising other unrelated parts of the client's experience. For instance, if a woman comes to counseling because of worry about her future and a sense of emptiness and dissatisfaction with her current life choices, it is important for her and the counselor to explore the effects of these concerns on her daily living and her relationships. Has this worry caused her to avoid friends, defer important financial planning, act angrily with her partner, drink more alcohol, and otherwise show a pattern of fairly wide disruption in her life activities? Or, is her concern more limited, perhaps causing some sleepless nights or loss of interest in some activities but little observable change in behavior or social relationships? By exploring the broader pattern of functioning, both the client and the counselor get a clearer picture of the nature and seriousness of the problem.

The fifth component of diagnosis is to explore the ways of solving the problem that the client has already tried before entering counseling. In American culture, counseling is often a "last resort" for people who have tried all of the other alternatives they could identify. Typically, clients have attempted a number of strategies on their own, and usually these solutions have failed or their success has been short-lived. The following situation serves to illustrate this point. The parents of a toddler who wakes up several times during the night may have attempted to let the child cry it out and then taken her into bed with them in an effort to teach her that nights are for sleeping. They may also have read a self-help book on the subject or sought the advice of their pediatrician. Only the failure of all of these strategies causes them to seek counseling. Counselor and client need to examine any prior attempts to cope so that fruitless strategies are not resurrected and so that the counselor can get a sense of the determination (or desperateness) of the client to get help. Some of the prior strategies may also have made the problem even harder to solve. In this case, attempting to let a toddler "cry it out" and then giving in and picking her up may have reinforced the child's learning that crying "works," in that it gets a parent to take her out of her crib at night. Thus, stopping the nighttime waking may be more difficult than it would have been had this strategy never been attempted. When counselors and clients understand the impact of the problem's history on the problem, strategies for change can be selected more prudently, and clients can be better prepared for the obstacles they may meet as they pursue these interventions.

The sixth component of diagnosis is an understanding of the strengths and coping skills of the client. Some of this information is revealed as the other five components are explored, but too often client strengths are not clearly identified. Knowing the strengths and coping skills of the client is important for several reasons. First, such information helps both the counselor and the client keep a balanced perspective. Even though the current difficulty is quite painful, the client does have resources to bring to its resolution and consciously focusing on them prevents a distorted view that the dysfunction represents the whole person. Because of their distress, clients often lose track of their positive characteristics and thereby may feel more overwhelmed by their problems than they really are. In addition, assessing strengths also helps the counselor remember that the client can take responsibility for himself or herself and should not be treated paternalistically. In other words, such a balance assists the counselor in avoiding either pitying the client or taking a parental role with the client. Finally, when several possible intervention strategies might be effective for a particular dysfunction, a knowledge of the client's strengths and coping skills can help counselor and client choose wisely. Related to a knowledge of personal strengths and coping skills is the need to be familiar with the support system available to the client. The central question is this: What significant others who are capable of positive interaction with the client are available during this stressful period? Clients without supportive persons in their lives may have substantially more difficulty in resolving their problems. In fact, in itself, the lack of such supports can be seen as a problem that may need to be addressed before the other counseling issues can be tackled.

Exploring all six components is important because typically two or three possible hypotheses about the client's discomfort emerge in the first and second stages of the counseling process. For example, a person's high anxiety in social situations may be related to low self-esteem, poor communication skills, a history of being demeaned by a parent or spouse, or a combination of these factors. By attending to all six components, counselor and client can begin to test out the accuracy of the possible explanations and rule out some and agree on others. Attending to all six components also helps the counselor avoid the mistake of deciding on a diagnosis too quickly and with too little data.

Counselors who work in community agencies are usually required to use the diagnostic system developed by the American Psychiatric Association published in the *Diagnostic and Statistical Manual of Mental Disorders* (3rd edition, revised), abbreviated as the DSM-III-R (American Psychiatric Association, 1987). A discussion of the strengths and weaknesses of this system is beyond the scope of this chapter. It is important to note, however, that the approach to diagnosis we present is not incompatible with the use of this system. Counselors using the DSM-III-R can still attend to all six components of the diagnostic process and work to make the hypothesis-testing and information-gathering process a joint enterprise.

Several other models for structuring the diagnostic process have been presented in the professional literature (e.g., Seay, 1978; Swenson, 1968). Arnold Lazarus (1976) published the best-known of these other models, which is labeled with the acronym BASIC ID. In his view, good diagnosis includes attention to behavior, affect, sensation, imagery, cognition, interpersonal relationships, and drugs. Lazarus contends that only with attention to all these client domains can the counselor accurately assess the problems and plan effective interventions. We encourage you to explore the views of these writers in greater detail to learn about other models of diagnosis.

TOOLS FOR EFFECTIVE DIAGNOSIS

Counselors rely heavily on verbal discussion with the client to arrive at a diagnosis. The process of facilitating self-disclosure and in-depth exploration, discussed in the Chapters 3 and 4, implies that verbal communication is the primary mode of analyzing and solving problems. However, counselors and clients have other tools available to them to assist in understanding the presenting problems. These include standardized tests, behavior rating measures, observations of the client in the natural setting, input from significant others, and role-playing exercises. Art, music, or play media may sometimes give counselor and client a deeper understanding of the issues. Art and play therapy techniques are standard diagnostic tools with children, but also have application with adolescents and adults. During the 1990s, computerized assessment programs for diagnosis will also become more common. This section briefly describes the array of tools available to assist in the diagnostic process. We encourage you to read sources such as Anastasi (1988) to acquire a full understanding of the use of these tools.

Counselors who are trained in the responsible use of standardized tests can find them a valuable resource for understanding clients, especially clients who present with complicated issues, who are not adept at verbal exploration of difficulties, or whose concerns are highly specific. Clients who have difficulty initiating conversation about embarrassing aspects of the problem may be more comfortable discussing such issues in the context of test scores or other "objective" information. Similarly, tests can reveal dimensions of a problem that a client never thought to bring up because he or she didn't see the connection to the presenting issue. Of course, no test is infallible, and all test data must be supplemented with other supporting evidence before any credibility can be given to such findings. When used responsibly, however, standardized tests can clarify confusing or hidden aspects of a problem. Professional guidelines for the responsible use of tests have been published by the American Psychological Association (1985) and the American Association for Counseling and Development (now the American Counseling Association) (1980).

Personality tests can be useful in situations where the concerns are long-standing or relate to negative feelings about self or difficulties in interpersonal relationships. Such tests compare the responses of the client to the responses of a norm group on a variety of dimensions of personality. Many personality inventories broadly assess intrapsychic functioning and attitudes toward others and thus help clarify the boundaries of the problem, the intensity of the feelings and the personal resources of the client. The most commonly used personality inventory in mental health settings is the *Minnesota Multiphasic Personality Inventory* (Hathaway & McKinley, 1967), or its revised version, MMPI-2 (Butcher, Dahlstrom, Graham, Tellegen & Kaemmer, 1989). Many other tests also examine multiple dimensions of personality. Some focus on identifying significant psychological dysfunction; others, such as the *Myers-Briggs Type Inventory* (Myers, 1975) and the *California Personality Inventory* (Gough, 1987), attend to personality style or the functioning of psychologically "normal" individuals experiencing situational problems. Those with extensive, specialized training may find projective personality tests, such as the *Thematic Apperception Test* (Murray, 1943) or the Rorschach (Rorschach, 1942), helpful in exploring client feelings and attitudes not fully available in conscious thought. Projective tests originated from the psychoanalytic tradition and assert that people project their unconscious difficulties onto their interpretation of ambiguous stimuli. For those who function in settings where the DSM-III-R system is used, the *Millon Clinical Multiaxial Inventory* (Millon, 1981) is a useful instrument for clarifying appropriate diagnostic categories within that classification.

There are also tests that center on a particular issue such as self-esteem, anxiety, or depression. These measures are appropriate when the general nature of the problem is fairly clear, but its seriousness or specific definition seems to allude the client and the counselor. One of the best-known of such instruments is the *Beck Depression Inventory* (Beck et al., 1961), which examines the intensity of sad and hopeless feelings and can help clarify depressed feelings that are obscured by anxiety or blocked from awareness by defenses. Such an instrument can also help determine the intensity of depression in a client who is not comfortable with conversing about such painful material. When examining self-esteem, for example, the *Tennessee Self-Concept Inventory* (Fitts, 1965) is widely used to explore more deeply the client's view of self. If used at intervals throughout the counseling process, these kinds of tests can serve as markers of client progress in specific domains.

When the client presents with a career or educational decision-making issue, interest inventories such as the *Strong Vocational Interest Inventory* (Hansen & Campbell, 1985) or *Holland's Self-Directed Search* (Holland, 1987) may be appropriate. They serve to articulate occupational and educational interests with a detail usually not possible through the interview process and to compare the client's pattern of interests with others in the same occupation. Interest inventories also function to familiarize clients with a broader range of occupational choices than they might be able to produce on their own.

There are several other tools available to help client and counselor make a good diagnosis. Included among these are observational checklists that involve either self-observation by the client or observation by someone else in the situations that provoke the problem. When the observation is of self, this procedure is also called self-monitoring. Structured checklists have been developed for special problems, such as school misbehavior, but individually developed models for self-observation or observation by others are commonly used in counseling as well. The critical component of these tools is a system for collecting data about the client's behavior that is easily understood by the person gathering the information. When the data is internal to the client, such as the number of self-depreciating thoughts, obviously only self-monitoring will be effective. Observational checklists and self-monitoring activities bring objectivity and specificity to client reports about negative experiences. Checklists are especially valuable when the difficulties seem centered on interpersonal interactions. For example, a client who finds himself getting into arguments with people in social situations but who doesn't understand the dynamics that lead to the disagreements may find it revealing to gather observational data to document the sequence of events. Similarly, when someone seeking a new job experiences repeated rejections after face-to-face job interviews, observation of interview behavior may unearth the problems better than any other diagnostic tool. (Practicality may dictate that observation be of a simulated interview rather than an actual one.) Also included on the list of useful tools are methods that involve input from significant others about client behavior. The value of such a tool is most apparent when a child is having difficulty, either in the classroom or at home. Because children do not have the cognitive or emotional maturity to observe their own behavior, feedback from a teacher or parent can be particularly helpful to the client and the counselor in diagnosing what's wrong. When input from significant others is included in the diagnostic data gathering, the counselor needs to be cautious about its interpretation, however. Most significant others will have difficulty being objective about the client and may inadvertently distort the data because of their personal relationship. Therefore, data from significant others should be supplemented with other, more objective information before a diagnosis is reached. Direct observation by the counselor is often helpful if it can be arranged.

Journals or diaries kept by clients can be another resource for defining the problem and for understanding the variations in the intensity of the problem from day to day. A journal may be a routine part of the client's life before counseling or can be "assigned" by the counselor for homework between sessions. Of course, in either case, the client must be completely willing to share such private writings with the counselor. For a client who is not interested in writing down thoughts and feelings, a tape-recorded journal may serve the same purpose.

Still other structured counseling techniques can be used to assess the presenting problem(s). As mentioned in Chapter 4, one of the most important of these tools is role playing during the counseling session. For example, a client who complains about not knowing how to make friends may be asked to role play with the counselor a contact with a new person he or she wants to get to know better. The counselor takes the role of the potential friend, and the client assumes his or her usual role. In this interaction, skill problems or negative feelings may come to the surface and be more readily discussed in the session. Later on, role plays can be used to test out the degree to which the client has learned new social skills or overcome negative feelings during such encounters.

Many other such techniques exist, and beginning counselors should be trained and supervised in their use before experimenting with them. However, the need for training and supervision should not deter beginning counselors from seeking to acquire these diagnostic tools because they are critical supplements to the verbal discussion in sessions. Of course, the responsible use of psychological tests demands careful training and a more cautious approach to experimentation.

PLACEMENT OF DIAGNOSIS IN THE COUNSELING PROCESS: RISKS AND OPPORTUNITIES

Essentially, the diagnostic process begins with the first contact between counselor and client. The fundamental motivation for seeking counseling is to get help in figuring out what is wrong, why independent efforts to change have not been successful, and to get the problem solved. Thus, the diagnostic process begins during the first session to some degree, but a concentrated focus on diagnosis at the beginning of counseling is a risky and often counterproductive strategy. Its risk is tied to the nature of the content to be discussed and the client's fears about how the counselor will respond. Specifically, because diagnosis involves both revelation of highly personal and painful information and attention to aspects of the situation that the client would rather avoid, it requires a trusting relationship between counselor and client. The client first needs to know that counseling is a safe environment where respect, caring, and understanding will be communicated. Only in such an atmosphere will the client be able to undertake the process of self-exploration essential to diagnostic accuracy. Counselors who focus immediately on finding a diagnosis are vulnerable to getting faulty or incomplete answers from clients to their questions. It is important to keep in mind that clients often fear that counselors will ridicule or judge them for their feelings and behaviors or will label them as sick or evil. Even if a client gives a counselor the benefit of the doubt about such harsh judgments, that client probably still needs to be reassured that his or her individual experience will be heard. A client often fears that he or she will be treated as "just another depressive" or a "typical example of low self-esteem" or labeled in some other way that diminishes uniqueness. These fears about the counselor need to be allayed before clients will feel free to divulge intimate information about their experience. Without time to establish trust, clients are more likely to be well defended and unable to process their own experience fully.

Diagnosis, then, clearly fits into the second stage of the counseling process as we define it. It is a focused and somewhat more structured activity within the deeper self-exploration of this stage, and like all such self-exploration, diagnosis can be threatening to the client. With full discussion of the problems inherent in covering all six components of diagnosis comes a confrontation with the reality that the client faces. Often that reality means an acknowledgment of the client's own role in the problem and a putting aside of the denials, rationalizations, or easy explanations that have served to make living with the situation more bearable in the short run, but are not compatible with real change. This deeper understanding is threatening because the old ways of thinking about the problem are relinquished before new ways of understanding it and coping with it are fully established. Clients sometimes drop out of counseling at this stage because the threat is too great. There is an old Irish proverb that helps explain this phenomenon: "The devil you know is better than the devil you don't know." In other words, however painful the current situation, there is almost always a way it could be worse. When clients commit to the deep exploration of the second stage and the diagnostic process, they are risking the banishment of the devil they know without any certainty that the path they choose will be better. Counselors who understand the threat will support the client's exploration in a way that acknowledges the risk and reminds the client of the possible gains from counseling.

The diagnostic process also presents an opportunity for the counselor to model effective problem solving. Clients often feel that their problems are insoluble and hopeless. During the days that preceded counseling, they probably experienced confusion and frustration whenever they tried to think clearly about their problems. The methodical process of exploring each aspect of the problem and the client's life situation will help bring clarity to the client. It also teaches the client another way to examine the issues. Furthermore, in watching the counselor proceed with the assumption that things can be improved or made easier to tolerate, the client gains hope that counseling will result in change.

INTAKE INTERVIEWS

In many college counseling centers, mental health agencies, and private counseling practices, a "tentative diagnosis" is reached during a separate intake interview conducted by an intake counselor. The intake interview usually involves a single meeting in which a counselor works with a client to gather information about the client's presenting problem, general life situation, history, and interpersonal functioning (Sommers-Flanagan & Sommers-Flanagan, 1993). The counselor and the client both understand that this session is separate from the regular counseling, and it is usually conducted by a person different from the regular counselor. One structure frequently used in intake interviews with severely disturbed persons is called the mental status examination. The mental status examination is a structured interview format that assesses current behavior and cognitive functioning. In this kind of intake interview, the counselor covers nine specific categories of information: client appearance, behavior/psychomotor activity, attitude toward the counselor, affect and mood, speech and thought, perceptual disturbances, orientation (to reality) and consciousness, memory and intelligence, and reliability and judgment (Sommers-Flanagan & Sommers-Flanagan, 1993).

The intake interview has become common because it can result in a more efficient use of staff resources and a better "match" between counselors and clients. When conducting intake interviews, counselors cover the same six components of diagnosis as they do in regular counseling sessions, but their attention must be highly focused. By necessity, intake interviews are more structured and directed by the counselor than are ordinary sessions. The use of an intake interview has its limitations, the most important of which is the need to discuss sensitive and painful information without time for trust and partnership between the participants to develop. (The difficulties with an immediate focus on diagnosis were discussed in detail earlier in this chapter.) In addition, clients can be taken aback by the possibility that they may have to "tell their story" again once a regular counselor is assigned. The counselor permanently assigned to a client does indeed read the intake report before the initial session with the client, but there are likely to be aspects of the report that the counselor will want to explore further with the client. Thus this worry has some basis in fact.

The intake interview has one other limitation. A directive intake procedure can leave clients with the impression that their proper role is to be rather passive, waiting for the counselor to ask questions and guide discussion. In other words, they may "learn" that counseling is a process akin to an appointment with a physician and thus take on the role of "patient." The client passivity that may result can undermine an effective partnership for problem solving. As a result, in an intake system, counselors have an added responsibility to educate the client about the mutuality of effective counseling and encourage the client's active involvement in the process. During intake interviews, counselors also have a special responsibility to make the client as comfortable as possible by attending to feelings and nonverbal behavior and using the active listening skills appropriate for the first stage of counseling. The questions and specific probes often required in the intake procedure must be built upon a foundation of these listening skills. (For a detailed discussion of methods for improving the effectiveness of intake interviews, see Sommers-Flanagan & Sommers-Flanagan 1993.)

Because the data collected in an intake interview may not be complete or accurate, counselors assigned for subsequent sessions ought to reexamine the content areas discussed in that interview once trust has been established and the client's self-exploration has begun. The diagnosis reached in an intake interview is always tentative, because the session is brief and the reliability of the information is unknown.

MISTAKES COUNSELORS OFTEN MAKE IN DIAGNOSIS

Counselors frequently make three errors during the diagnostic process. The first of these is the assumption that the difficulties the client is experiencing are caused by psychological, interpersonal, or social factors. Clients sometimes experience distress because of medical dysfunctions, or their psychological pain is complicated by medical problems. In these cases, the sadness, irritability, or anxiety can only be treated properly if the physical condition is remedied. As a routine part of the diagnostic process, counselors need to explore the client's medical history. Clients who have not had a physical examination by their medical doctor since the onset of their distress should be encouraged to schedule such an appointment. The need for a medical check-up is especially important when the distress is persistent, severe, or of sudden onset. A client's distress may be related to a medication that has been prescribed for an unrelated medical problem. Clients who are not familiar with the side effects of their medication ought to discuss this possibility with their physician. Most of the time, of course, the sadness or irritability a client is experiencing is not related to any medical problem, and counseling interventions will be necessary to help the client change. However, since the medical problems that can cause emotional changes are often serious and sometimes life-threatening, counselors should encourage all clients to see their physician in order to rule out physiological explanations for their difficulties.

The second mistake counselors sometimes make is thinking that there is only one diagnosis for a client's problems. Typically, clients come to counseling with more than one difficulty or more than one domain in which their functioning is somewhat compromised. Consider the following case.

Case: Howard

Howard enters counseling with the statement that "he's at his wit's end." Exploration of the problem reveals that he has recently been laid off from work, that his 27-year-old son has recently been diagnosed as HIV-positive, and that he is having frequent panic attacks and is becoming afraid to leave his home. Howard goes on to say that he and his wife have been unable to give each other support in their worry about their son's illness, that they have been arguing often, and that he has lost interest in sex.

Discussion

In this case, each difficulty ought to be fully explored, and the relationship between the problems examined. Real grief about his son may be underlying Howard's anxiety and distance from his wife. Financial worries may be preventing him from attending to his grief, his partner, or his job search. His loss of interest in sex may be a consequence of the decline in emotional closeness with his wife or may be a symptom of an unrelated medical condition.

Questions for Further Thought

1. Imagine yourself in Howard's situation. How would you feel and what might you expect from your counselor? Share your response with a partner.
2. With the same partner, identify several other diagnostic possibilities for this case.
3. Referring back to the six components of diagnosis, identify other diagnostic information you would need to assess Howard's counseling needs accurately.

No single diagnosis is likely to account for all of Howard's concerns. Counselors who believe that there can be one, overriding diagnosis for complex situations such as this one risk oversimplifying the situation and thereby, neglecting intervention in some important aspects of the problem.

The third mistake counselors may make is treating a diagnosis as though it were unchangeable and absolute, when, of course, a diagnosis is always tentative to some degree and always open to new information. The process of exploring and making changes in one area may bring into the client's awareness other issues that he or she had no memory of before that point. Such forgotten or repressed issues may be highly painful; examples are a history of physical abuse, an instance of date rape, and a highly disturbing combat experience. This painful information becomes available to the client when trust in the counselor has been tested out a number of times and when the client begins to feel stronger in other ways. When new disclosures are made, the counselor and the client need to reassess the diagnosis and amend it if necessary. If counseling is at the third stage and plans for change have been implemented, then both parties need to decide whether the interventions are still appropriate. Often, additional interventions are needed.

MAKING REFERRALS WHEN COUNSELING ISN'T APPROPRIATE

Because of the social stigma attached to seeking counseling, it is more common for people to seek out medical care for psychological and interpersonal difficulties or to try to self-treat such distress than to make an appointment with a counselor. Sometimes, however, clients define their problems as psychological when, in fact, other factors are at the foundation. It is during the diagnostic process that the appropriateness or inappropriateness of counseling for dealing with the problem becomes fully apparent. As mentioned above, clients may misinterpret symptoms of physiological problems as psychological. For example, aging clients who are experiencing side effects of medications prescribed for them do not need counseling for their symptoms; they would benefit most from changes in their medical care. Some clients come to counseling with problems that are better served by other helping professionals. For example, a woman is experiencing great distress because of the disrepair of her apartment in a rent-controlled building. She feels anxious and very angry. Exploration of the problem reveals that she has a good support system, flexible coping skills, and no history of any dysfunctional behavior. In other words, in spite of her feelings, she is still functioning very well in the activities of her day-to-day life. In this case, referral to social service agencies or legal practitioners who can help her respond to the landlord's failure to act responsibly is really more appropriate than counseling as the primary avenue of intervention. Clearly, the counselor can play a role in providing support and should maintain contact with the client, but the most therapeutic intervention is to get the client in touch with those who have the expertise to assist her.

A thorough diagnosis that attends to all six components will provide counselor and client with the data necessary to make the determination about whether a referral is necessary. Counselors should not assume that counseling is always the best or even an appropriate response to the client's distress; that determination awaits evidence. However, once a client does make contact with a counselor, that counselor has the responsibility to carefully assess what is wrong and provide the client with referral sources if counseling is not appropriate. If the client's problem is beyond the counselor's expertise, then a referral to a counselor whose skills better match the needs of the client is appropriate. One of the ethical standards of the profession is to work within the limits of one's professional competence.

SUMMARY

Diagnosis in counseling is the activity of systematically defining the difficulties that caused the client to seek counseling. Diagnosis is not something the counselor does to the client, but rather an information-gathering and hypothesis-testing process carried out in collaboration with the client. Diagnosis can be somewhat threatening to the client since it necessarily includes deep exploration of sensitive material and since it implies risking change. Consequently, diagnosis is best conducted only after a trusting relationship is well underway and the client understands the mutuality of the counseling process. Counselors can anticipate some resistance to diagnosis from clients, because of the confrontation with sensitive material. When resistance occurs, attention to empathic understanding and active listening skills will usually serve to diminish it.

There are six components to a good diagnosis: an definition of the boundaries of the problem; a mutual understanding of the factors that maintain and lessen the problem; an understanding of the intensity of the feelings surrounding it; a definition of the degree to which the presenting problem is affecting functioning in other parts of the client's life; an exploration of the attempts to solve the problem undertaken prior to counseling; and an identification of the strengths and coping skills of the client.

When there are multiple problems, a counselor explores each difficulty with the same thoroughness. The counselor should also be open to the possibility that counseling is not the best intervention or that the client needs a different counselor. In such cases, the client should be referred to other helping professionals better equipped to address the client's concerns.

REFERENCES

American Association for Counseling and Development. (1980). *Responsibilities of users of standardized tests.* Alexandria, VA: Author.

American Psychological Association. (1985). *Standards for educational and psychological tests.* Washington, DC: Author.

American Psychiatric Association. (1987). *Diagnostic and statistical manual of mental disorders* (3rd ed., rev.). Washington, DC: Author.

Anastasi, A. (1988). *Psychological testing* (6th ed.). New York: Macmillan.

Beck, A., Ward, C. H., Mendelson, M., Mock, J., & Erbaugh, J. (1961). An inventory for measuring depression. *Archives of General Psychiatry, 4,* 53–63.

Brammer, L. M., Shostrom, E. L., & Abrego, P. J. (1989). *Therapeutic psychology: Fundamentals of counseling and psychotherapy* (5th ed.). Englewood Cliffs, NJ: Prentice-Hall.

Butcher, J. N., Dahlstrom, W. C., Graham, J. R., Tellegen, A., & Kaemmer, B. (1989). *MMPI-2 manual for administration and scoring.* Minneapolis, MN: University of Minnesota Press.

Fitts, W. H. (1965). *Manual: Tennessee self-concept inventory.* Nashville, TN: Counselor Recordings and Tests.

Gough, H. G. (1987). *California personality inventory.* Palo Alto: Consulting Psychologists Press.

Hansen, J. I., & Campbell, D. (1985). *The Strong manual.* Palo Alto, CA: Consulting Psychologists Press.

Hathaway, S. R., & McKinley, J. C. (1967). *Minnesota multiphasic personality inventory manual.* New York: Psychological Corporation.

Holland, J. L. (1987). *The self-directed search professional manual.* Odessa, FL: Psychological Assessment Resources.

Lazarus, A. A. (1976). *Multimodal behavior therapy.* New York: Springer.

Millon, T. (1981). *Disorders of personality: DSM III, Axis II.* New York: Wiley.

Murray, H. A. (1943). *Thematic apperception test.* Cambridge, MA: Harvard University Press.

Myers, I. B. (1975). *Manual: The Myers-Briggs type indicator.* Palo Alto, CA: Consulting Psychologists Press.

Rogers, C. R. (1951). *Client-centered psychotherapy.* Boston: Houghton Mifflin.

Rorschach, H. (1942). *Psychodiagnostics: A diagnostic test based on perception* (4th ed). New York: Grune & Stratton.

Seay, T. A. (1978). *Systematic eclectic therapy.* Jonesboro, TN: Pilgrimage Press.

Shaffer, W. F. (1986). Diagnosis as a sham and a reality. *Journal of Counseling and Development, 64,* 612–613.

Sommers-Flanagan, J., & Sommers-Flanagan, R. (1993). *Foundations of therapeutic interviewing.* Needham Heights, MA: Allyn and Bacon.

Stiver, I. (1986). The meaning of care: Reframing treatment models for women. *Psychotherapy, 2,* 221–226.

Swenson, C. H. (1968). *An approach to case conceptualization.* Boston: Houghton Mifflin.

CHAPTER 6

STRUCTURING, LEADING, AND QUESTIONING TECHNIQUES

The process of counseling has been described in previous chapters as a sequence of stages in which the focus shifts from initial disclosure of concerns to deeper exploration to plans of action. These stages may all occur within a single session, but more commonly the process develops over two or more sessions. We have urged the counselor to pay special attention to building a good relationship with the client early in the counseling process. Later, because of continuity of this positive relationship, its maintenance will require less time, and the trust that has been built can make it easier for the client to look at deeper and more threatening issues. Commitment to action may require the counselor to plan actively with the client and to reinforce the client's efforts to change.

This chapter considers the moment-to-moment decisions that a counselor makes each time he or she responds to a client. In the instant just after a client has completed a communication to the counselor (verbally or otherwise), the counselor has a whole universe of possible responses available. In a few seconds, the counselor selects a response and, in so doing, influences the subsequent response of the client and ultimately the direction counseling will take. However, as Ivey (1988) has stated, there is no one "correct" response at any given instant, but rather many alternative responses that may move the client toward a sense of capability.

A similar-appearing process occurs in ordinary conversation, of course. However, because counseling is a dialogue with a purpose and the counselor is responsible for facilitating positive progress, individual responses are more significant. If one wishes to improve one's counseling skills, it is often necessary to study the timing, content, and effect of specific responses. Supervised practice in counseling provides counselors-in-training with feedback on how specific responses contributed to or interrupted a client's progress. This chapter will help with the formulation of specific counseling responses and will provide some structure for analyzing the counseling process bit by bit.

STRUCTURING

Structuring is any statement by the counselor that lets the client know what to expect of the process and outcomes of counseling. It may address "the nature, limits, conditions, and goals" of counseling (Brammer, Shostrum, & Abrego, 1989, p. 119). Structuring helps to keep the conversation purposeful. According to Corey (1991), "All the approaches (to counseling) are in basic agreement on the need for some type of structure in the counseling experience, although they disagree over its nature and degree" (p. 433).

Many clients arrive at a counselor's office having no idea what to expect of counseling. Others arrive with unreasonable or inappropriate expectations. Patterson (1974) observed that many clients not experienced with counseling think of counselors as experts who will give them advice and solve their problems. Some of a counselor's earliest statements to a client will suggest how the client might participate and what the counselor will contribute to the conversation. Still other clients lose momentum during counseling and need help in maintaining the motivation to work on their concerns or to move to a new stage. Therefore, it may be necessary for the counselor to return to structuring at various times throughout counseling.

The counselor must decide how much structuring to use and when to use it. Too little structuring contributes to the possibility of rambling, unfocused interaction that is lacking in concreteness and is unproductive; another possible outcome is that the client will present an initial concern and then withdraw to await the counselor's solution. Too much structuring sounds preachy and may seem to the client like scolding for not participating "properly." It is also possible, especially with clients who have previously experienced counseling, that poorly timed efforts at structuring may interrupt important client thought processes. Formative statements that refocus the working agenda when necessary are better than protracted speeches that describe the counseling process. These formative statements are timed so as to set the initial working agenda or to maintain the momentum of the client's work on his or her concerns.

Structuring at the Beginning of Counseling

During the first couple of sessions, the pattern of how the client and counselor will work together is established. With adolescent and adult clients, the counselor will usually initiate specific discussion about confidentiality, length and frequency of counseling sessions, projected duration of the counseling, client and counselor responsibilities, and possible outcomes. With children, there is often discussion of action limits as well (e.g., "You may not break the equipment or hit the counselor"). Also, wording of the other elements of structure must be adjusted to the client's ability to comprehend.

Confidentiality frees clients to talk about personal material without the fear that what they say will be repeated to others who might use the information against them or think less well of them as a result. Discussing confidentiality early in the counseling process and any other time it is introduced by the client is therefore important. With a client who is known to have experienced counseling previously, the counselor may begin by asking about the client's understanding of confidentiality and then elaborating as necessary. With younger clients and those who may not have had previous exposure, the counselor can make a statement worded something like this: "I want you to know that the things we talk about together are private and I will not discuss them with anyone else without your permission. There are also a couple of exceptions you should know about. If you talk about harming yourself or someone else, or if you talk about being in danger, I may have to involve someone else to help." It is a good idea for trainees working under supervision to tell the client that a supervisor may also hear some of the sessions and that the supervisor will also respect the client's privacy (Sommers-Flanagan & Sommers-Flanagan, 1993). Although confidentiality is often taken for granted by counselors, a client who wishes to share highly personal information will be assured by explicit statements that his or her story will not be treated casually.

Discussion of the time frame for counseling tells the client what to expect and provides a perspective for how he or she will choose to use counseling. A statement such as "We will have forty-five minutes to talk together today" at the beginning of the first session lets the client know it is not going to be a hurried contact. If it is determined that additional counseling is needed, the client should be told that a specific time period (time and day) will be reserved for future contacts. This allows the client to look forward to future sessions and to anticipate how he or she will use them, while at the same time communicating the expectation that the client will be independent and not seek daily contact. Many counselors also set a target for completing counseling after hearing what the client wants to work on. This might be framed like this: "Let's plan to meet for about six sessions and then take stock of where you are. Often people are able to sort out concerns like yours in about that amount of time."

The most complex structuring skills are those that convey to the client what to expect of the counseling process itself. Wallace (1986) says that "clients should leave the initial interview with a clear understanding that therapy is a collaborative effort, that they are expected to be active participants in the therapeutic process, and that, although there is certainly reason to hope, there is no promise of success" (p. 71). The client's part of the collaborative effort includes sharing important and sometimes threatening material, making a sincere commitment to changing those things that can lead to problem resolution and growth, working with the counselor to develop alternatives, and taking action as plans develop. The counselor's role is to help the client tell his or her story, to facilitate the client's understanding of his or her feelings, motives, and behaviors, to help the client in reframing aspects of his or her perspective on reality, and to build action plans with the client. Although the client may learn these complementary roles through experience with counseling, it is useful if the counselor starts the process in the first session with statements like this: "I can see that you are very troubled by what is happening in your relationship with your son, and I will try very hard to help you understand where the problem is coming from and what you might be able to change." The key element here is that the client must clearly understand that he or she will be left in charge of his or her own life. The counselor helps to enhance the client's coping skills but does not take over solving the problems presented.

A well-structured first counseling session begins the initial-disclosure process. It also helps the client understand what to expect from counseling (and what not to expect), how he or she should participate, and something of the time frame involved.

Structuring Later in the Counseling Process

When structuring is used later in the counseling process, its purpose is to remind the client of the nature of the process and to reinforce the appropriate roles of the participants. It may also move counseling forward to a new stage. Examples of such structuring may be drawn from Carl Rogers' (1965) classic demonstration film with a client named Gloria. Gloria is trying to decide whether or not to tell her young daughter about her sexual behavior. Would it be more harmful to the relationship to tell her daughter about having sex or to lie to her about sex? She directly asks Dr. Rogers for his opinion. He responds, "I surely wish I could give you an answer, because an answer is what you want. . . . But that would be no damn good." He goes on to make clear that there are risks involved with both courses of action and that only Gloria can decide which risks she is more willing to take. He can help her clarify the consequences of each of the choices, but she must make the choice because she must live with the consequences. Later, the counseling session seems to bog down. Gloria has carefully explored what appear to be the only two courses of action available to her, but she seems unprepared to take any action. Dr. Rogers gently moves her to the action stage by suggesting that he believes she knows what she wants to do. He senses that she places a higher value on being honest than appearing "pure." Gloria affirms his perception and begins thinking about how to proceed.

LEADING

A counselor is typically confronted every few seconds during a counseling session with a choice about how to respond to the content and affect of what the client has just said. The potential range of choices is infinite, and there are usually several responses that have positive potential for advancing the counseling. Other responses may not contribute much to progress, and still others may actually delay progress or disrupt the counseling process altogether. Many potential counselor responses—paraphrase, reflection of feeling, probing, and interpretation—were discussed in Chapters 2, 3, and 4. The purpose of this section is to assist counselors in the moment-to-moment selection of specific responses based on anticipation of client readiness.

Robinson (1950) coined the term *leading* to describe the counselor's selection of a response that anticipates the client's readiness to benefit from a particular kind of response. As discussed in Chapter 4 (on in-depth exploration), responses that include elements of confrontation and interpretation may be valid but nevertheless destructive to the purposes of counseling if introduced before the client is ready to accept and integrate the information contained in them. The concept of leading includes the proposition that there is "a critical region, just ahead of but not too far ahead of the client, where therapy takes place most efficiently" (Martin, 1989, p. 25).

Robinson (1950) used a football analogy to describe what he meant by leading. When throwing the football, the passer anticipates where the receiver will be when the ball arrives and throws the ball out ahead of the receiver (leads the receiver) so that the ball and the receiver arrive at the same place at the same time. Analogously, Robinson advised the counselor to estimate where the client is going next and to formulate a response that will intersect with the client's path and further his or her progress. The analogy can be extended to include the concept of *length of lead*. If a counselor underestimates the pace of a client's progress, he or she forces the client to slow down and react to a statement that from the client's viewpoint needs no further work—much like a pass receiver must slow his pace or retrace his footsteps to catch a pass that is underthrown. If a counselor overestimates a client's pace, he or she may make a statement that is beyond the client's ability to comprehend and internalize, and the client may become confused and defensive. Progress is then impeded—much as in the case of a receiver who just can't make it to the football because it has been overthrown. When trying to estimate a client's *leading edge*, a counselor includes everything that has occurred in his or her experience with the client that the client intended as signs. "This includes verbal and nonverbal cues as well as material that the client has 'put on the table' and assumes you know from previous discussions" (Martin, 1989, p. 26).

In this chapter, we use the word *lead* to mean any response that is based on the counselor's estimate of the likelihood of making contact with the client's next awareness. We do not intend to suggest that the counselor leads the client as one might lead a small child by the hand through a crowded store. Nevertheless, the counselor does select responses (overtly or covertly) that he or she thinks will encourage the client to continue moving ahead in the process. "The term *lead* means a teamlike working together in which the counselor's remarks seem to clients to state the next point they are willing to accept" (Shertzer & Stone, 1980, p. 272).

Continuum of Lead

Some responses stay very close to what the client has just said and introduce very little of the counselor's thoughts and feelings; other responses take a rather large step from the client's most recent statement and introduce a considerable amount of the counselor's perspective (Hansen, Stevic, & Warner, 1982). Robinson (1950) suggested that it is useful to arrange all types of response on a continuum from least leading to most leading, thereby creating reference points that facilitate the choice of slightly more or less leading responses as they are needed in counseling sessions. Such a continuum provides a convenient system for comparing the impact of different counselor responses.

So many different terms have been used over the years to describe various kinds of counselor response that it is not possible to include all of them. Table 6.1 illustrates the usefulness of placing responses on a continuum; it is based on Robinson's early work and uses his terminology and order of techniques. Benjamin (1987) includes a very similar continuum in his useful description of the helping process. Consider the terminology used in Chapters 3, 4, and 5 of this book as it relates to the continuum. For example, primary empathy and advanced empathy are probably best categorized as increasingly leading responses under restatement and clarification, respectively. Constructive confrontation fits best under the rejection/persuasion moniker.

It is important to realize that all categories of response indicated on the continuum of lead are useful and appropriate at some time in the counseling process; none should be thought of as "good" or "bad," "right" or "wrong" (Benjamin, 1987). The degree of lead—that is, the distance the counselor moves toward the most leading end of the continuum—will depend upon the client's readiness, the kind of concern being discussed, and the predisposition of the counselor. If a client seems particularly defensive, as in the case of a student referred to a school counselor after starting a fight in the cafeteria, the counselor is well advised to take it easy and try to learn the client's view of the incident. On the other hand, a student who is seeking information and guidance about the selection of a college may be well served by counselor-initiated ideas and information and frustrated by a lot of restatements of his or her need.

It is the nature of some counselors to listen a lot and be very cautious about reaching *any* quick conclusions about another person. Such counselors maintain minimal lead in working with clients, using a lot of restatement and clarification responses and occasionally remaining silent. Other counselors believe that they can size up clients fairly quickly, and they become impatient

Table 6.1 Continuum of Lead

Least leading response	Silence	When the counselor makes no verbal response at all, the client ordinarily feels some pressure to continue and chooses how to continue with minimum input from the counselor.
	Acceptance	The counselor simply acknowledges the client's previous statement with a response such as "Yes" or "Uh-huh." The client is verbally encouraged to continue, but without content stimulus from the counselor.
	Restatement (paraphrase)	The counselor restates the client's verbalization, including both content and affect, using nearly the same wording. The client is prompted to reexamine what has been said.
	Clarification	The counselor states the meaning of the client's statement in his or her own words, seeking to clarify the client's meaning. Sometimes elements of several of the client's statements are brought into a single response. The counselor's ability to perceive accurately and communicate correctly is important, and the client must test the "fit" of the counselor's lead.
	Approval (affirmation)	The counselor affirms the correctness of information or encourages the client's efforts at self-determination: "That's good new information," or "You seem to be gaining more control." The client may follow up with further exploration as he or she sees fit.
	General lead	The counselor directs the client to talk more about a specific subject with a statement such as "Tell me what you mean," or "Please say some more about that." The client is expected to follow the counselor's suggestion.
	Interpretation	The counselor uses psychodiagnostic principles to suggest sources of the client's stress or explanations for the client's motivation and behavior. The counselor's statements are presented as hypotheses, and the client is presented with potentially new ways of seeing self.
	Rejection (persuasion)	The counselor tries to reverse the client's behavior or perceptions by actively advising different behavior or suggesting different understandings of life events than those presented by the client.

(continued)

	Reassurance	The counselor states that, in his or her *judgment,* the client's concern is not unusual and that people with similar problems have succeeded in overcoming them. The client may feel that the reassurance is supportive but may also feel that his or her problem is being discounted by the counselor as unimportant.
Most leading response	Introducing new information or a new idea	The counselor moves away from the information or client's last statement and prompts the client to consider new material.

Adapted from F. P. Robinson, *Principles and procedures of student counseling.* New York: Harper & Brothers, 1950.

with listening to further iterations of a problem once they think they understand it. Such counselors are likely to share their perceptions with clients earlier and thereby maintain a longer lead. These counselors will make rapid progress with clients whose defenses are moderate and who have good cognitive skills, but their leading may be threatening or simply confusing to clients who lack ego strength or cognitive skills. To some degree, a counselor's preference concerning the use of leads is influenced by the theoretical training he or she has had. (See Chapter 12 for further development of the relationship between theoretical systems and preferences for leading.)

The counselor's decision about length of lead is related to effectiveness and efficiency. On the one hand, it is important to keep length of lead modest enough to stay in contact with the client; on the other hand, the longer the lead that can be achieved while still maintaining contact, the more rapid the progress. Minimum distance leads (e.g., paraphrases) are least threatening to the client because they do not pull him or her from a current level of understanding to some surprising hypothesis about self. However, there is efficiency in pulling a client toward a new perspective if his or her defenses will permit the assimilation of the new material; a longer lead (e.g., interpretation) may result in quicker development of insight if the client is ready. It is generally recommended that inexperienced counselors make conservative judgments (short length of lead) while gaining experience, because excessive numbers of longer leads can frighten clients and preclude effectiveness of counseling. Longer leads become safer once diagnostic abilities have been honed through experience.

Leading and the Stages of Counseling

Counselor responses usually become increasingly leading as counseling proceeds through the three stages. In the early phase of counseling, when relationship building is of prime importance, the counselor's responses tend toward the least leading end of the continuum. In-depth exploration is prompted by responses in mid-range. Commitment to action often requires some counselor-initiated reinforcement as well as counselor-supplied information—both examples of responses toward the most leading end of the continuum.

A review of the material in Chapter 3 will reveal that the counselor uses mostly responses from the least leading end of the continuum during the initial-disclosure stage of counseling. Restatements or paraphrases are commonly used to communicate primary empathy. The following example involved a worried 8-year-old girl.

> CLIENT: I don't know what I'm going to do. My mother doesn't live here any more, and my grandma says she is not going to be my mom.
>
> COUNSELOR: You're scared that there won't be anyone to take care of you, and you are too little to take care of yourself.

The counselor accurately perceived the emotion and the basis for that emotion. Such a statement is said to be *interchangeable* with the client's statement.

By restating the material for the client, the counselor showed understanding and encouraged the client to continue with the disclosure process. The client went on to say that her mother had left town with a man after leaving her with her grandmother. Although the grandmother had been providing basic care, she had made it clear to the girl that she wasn't very interested in getting back into the business of raising a child. Using restatement, acceptance, and silence, the counselor was able to acquire a fairly complete picture of the problem from the child's point of view.

Because the child could not be expected to arrive at her own solution to the problem of who was going to take care of her, counseling did not proceed to the second and third stages. Instead, the grandmother was invited for consultation, and the counselor helped her understand that the client took her statements about not wanting to have another child to care for as a threat that she too would abandon the child. Fortunately, the grandmother was committed to providing care for as long as the daughter was gone, and she was able to be more supportive with the child, saving her anger to express to the daughter when she could do so.

The second stage of counseling, as described in Chapter 4, promotes in-depth exploration by the client. Therefore, responses toward the middle of the continuum are likely to be most useful.

Advanced empathy, as noted earlier, would probably be considered a clarification response by Robinson. With advanced empathy, the counselor uses more psychodiagnostic skills in understanding the meaning of the client's statements and provides a response that moves to a point of clarification based on the signs but going beyond what the client has actually said. In the following example the client is a 27-year-old woman.

> CLIENT: I got tired of just sitting around so I went down to the bar to see if Frank was there. I almost hoped he wouldn't be there, but I had to get out.
>
> COUNSELOR: It sounds like you didn't want to stay home alone, and you didn't want to see Frank either. You were pretty desperate just to have some human contact.
>
> CLIENT: Yeah. When I sit around alone for a while, I begin to wonder if I'm okay. I don't seem to know what I want. I know Frank's a bum, but at least when he's around I don't feel so lost.
>
> COUNSELOR: When Frank is around, you don't feel quite so incomplete as when you are alone, and that is less frightening even though you know Frank has characteristics that are problematic to you and others.

In this dialogue, each of the counselor's responses added perspective and clarity to the client's statement. In the first instance, the counselor magnified the affect of the client's statement by responding to the sense of desperation. In the second instance, the counselor labeled the sense of incompleteness that was at the root of the desperate feeling. This led to further in-depth exploration in which the client affirmed that the feeling she had when she was alone caused her to accept the company of virtually anyone who would spend time with her. This had led to a series of unsatisfying and abusive relationships with men whom she had met in bars. The responses in this example are referred to as *additive* because they add perspective from the counselor's assessment of the client's concern. Long-term counseling was required for this client to begin to feel a sense of personal integrity and worth that enabled her to begin to select more satisfying company and feel comfortable when alone.

Interpretation is also used heavily to promote in-depth exploration. Interpretation can be seen on the continuum of leads as being beyond clarification statements in a position where the counselor is adding even more diagnostic content to his or her response. There is no hard-and-fast rule for distinguishing strongly additive empathy statements from mild interpretations. An example of an interpretive response follows. The client is a 15-year-old boy.

> CLIENT: I don't know why everyone keeps bugging me about getting grades to get into college. I can always pump gas. *(The client smiles as he makes the last statement and assumes the pose of holding the nozzle of a gas pump.)*

COUNSELOR: Yeah, you could really get even with your dad by messing up in school.

As is characteristic of interpretive responses, this response seems harsh and irrelevant if one doesn't know the background of the client. The client's father was a self-made businessman who owned and operated four gas stations. The father had placed a great deal of pressure on his son "to make something of himself," that is, to become an educated man who would not have to do work his father considered demeaning. The son resented this pressure, and his gas pump pose was a passive-aggressive cliché he used to anger his father. The client saw nothing demeaning about pumping gas. On the other hand, he had little awareness that his motivation to work at that level in the family business came more from rejection of his father's wishes than from any real interest in the business. What the counselor said was an attempt to get the client to consider his motive. By definition, interpretation goes beyond the client's present conceptualization of a problem, and it will therefore always present the client with a new idea.

Techniques closer to the most leading end of the continuum are used more frequently in the third stage of counseling—commitment to action. By this stage, the counselor understands the nature of the client's concerns and has a good sense of the client's goals. With that information as a basis, active encouragement through persuasion and reassurance may be needed to help the client act on what he or she wants to do. At this stage of counseling, the counselor may appropriately introduce new ideas—often information about alternatives that will assist the client by expanding his or her awareness of available options. Such responses are sometimes called *initiating* responses. (See Chapter 7 for a more specific discussion of counseling skills to promote client actions.)

The following principles serve to relate the degree of lead to the stage of counseling:

1. Minimum-lead responses are excellent for relationship building.
2. Minimum-lead responses are low-risk responses, because they do not frighten the client with startling new perspectives.
3. Responses toward the center of the continuum increase the client's in-depth exploration.
4. Responses toward the center of the continuum can be somewhat threatening and may produce defensive reactions if used prematurely.
5. Maximum-lead responses are very directive, reinforcing specific behaviors that relate to the client's goals.

6. Maximum-lead responses are likely to seem irrelevant to the client unless the counselor has taken time to know the client and his or her goals.

THE USE OF QUESTIONS IN COUNSELING

Questions, depending on their formulation, can embody a minimum, middle, or maximum degree of leading. Good questions are useful in opening new aspects of the client's situation for discussion (such as history, strengths, or prior attempts at alleviating the problem), in clarifying vague or conflicting comments from the client, or in focusing the client's attention on specific thoughts, feelings, or behaviors related to the problem. Well-structured and well-planned questions move the counseling process ahead and promote insight and motivation to change, but poorly worded or mistimed questions interfere with that process. Well-planned questions follow from the client's prior statements without an abrupt change of topic or tone. For example, if a client has been discussing his ambivalence about seeking a more satisfying job and risking his current security, any question about current stress in his marriage or his worries about his children must be connected to the client's concern and posed in such a way that the client can recognize the connection. In addition, effective questions should deal with content that the client can comfortably handle, given the level of trust established and the depth of self-understanding achieved at the point at which the question is asked. Questions that "go deep into the client's defenses" or touch on painful material must be planned to generate the least amount of resistance possible. Sometimes, a question is poorly planned because it rushes the client to define the problems too quickly or hurries the process of identifying possible solutions to the issue. A counselor who hears himself or herself asking questions beginning with "Have you tried . . ." or "Do you think your problem is . . ." should examine whether they represent such a hurrying of the client.

The wording of questions is equally important. The general rule is that questions should be phrased so that they encourage the client to answer them in long phrases or full sentences. Questions that have yes/no answers are less useful and may result in a pattern in which the counselor needs to respond again after the client's one-word answer. Consider the following interactions:

Interaction A

CLIENT: I feel so uncomfortable with my family lately. I'd rather be alone or at the office working.

COUNSELOR: What feelings are included in that discomfort?

CLIENT: Mostly I feel like such a failure, and I think that they can't really love or respect me.

Interaction B

CLIENT: I feel so uncomfortable with my family lately. I'd rather be alone or at the office working.
COUNSELOR: Do you feel distant with them?
CLIENT: Yes, often.

The second interaction generated little new information from the client and set up a pattern that required the counselor to ask something more to get the client to keep disclosing. For this reason, yes/no questions are also called *closed-ended questions.* They tend to close off discussion of an issue.

Another risk of yes/no questions is apparent from a careful examination of the two interactions. Forming such a question required the counselor to guess which feelings might be inside the client in order to name one to ask about. The client's discomfort might have been anger, sadness, distance, worry, or several other feelings. The process of identifying the most likely client feeling in order to form a question is similar to the process of constructing an interchangeable response. However, a question more narrowly directs the client's response, and the consequence of erroneously identifying a feeling is to more or less force the client to talk about that feeling, whether germane or not. In Interaction B, "distant" is not a highly affective word and does not promote much response. By asking the question in a more open-ended way, as demonstrated in Interaction A, the counselor can avoid the pitfall of guessing at the client's internal experience.

For similar reasons, questions that are phrased in an either/or way are not usually helpful. Asking the client in the above interactions whether he felt distant or angry at his family would simply have represented two guesses at his feelings, both of which could have been wrong. And the client might have responded with a one-word answer: "Distant," "Angry," or "Neither."

Counselors also sometimes fall into the habit of using questions that are not really questions. The following example with a fifth-grade girl who is resisting efforts to help her adjust to her parents' move to another state provides an illustration:

CLIENT: It's just not fair that I have no say in this decision. I want to stay and live with my friend, Lucy, but my dad won't even consider it.
COUNSELOR: Don't you think you'd miss your parents after a while?
CLIENT: No. Besides, I could visit them at school vacations.

In this case, the counselor's question was really a statement of her opinion about the girl's reaction to such an arrangement, even though it was worded as a question. The counselor's hidden agenda may have been to get the girl to see another point of view, but it backfired when approached in this fashion. A question that begins "Don't you think . . ." or "Don't you feel . . ." probably should be rephrased. In this example, an open-ended question such as "How do you think it would work out if you lived with Lucy?" would show more respect for the client's viewpoint and help her explore the reality of such a situation.

SUMMARY

This chapter considered techniques of counseling to be specific types of statements that are used with a client. Structuring statements are designed to help the client learn to use counseling and learn the role and responsibilities of client and counselor. Leading responses are statements that prompt the client to examine his or her own statements. The counselor attempts to judge what degree of leading the client is ready for and to provide a response to the client's most recent statement that will move the process forward. Typically, in the early stages of counseling, leading is at the minimum end of the continuum. In-depth exploration is best promoted by responses toward the center of the continuum, and commitment to action may involve responses at the most leading end, including use of new information and reinforcement of selected behaviors. Questions can occur at any time in the counseling process. If they are open-ended and well-timed, they promote deeper client self-exploration. However, poorly formulated or badly timed questions interrupt client progress.

REFERENCES

Benjamin, A. (1987). *The helping interview with case illustrations.* Boston: Houghton Mifflin.

Brammer, L. M., Shostrum, E. L., & Abrego, P. J. (1989). *Therapeutic psychology* (5th ed.). Englewood Cliffs, NJ: Prentice-Hall.

Corey, G. (1991). *Theory and practice of counseling and psychotherapy* (4th ed.). Pacific Grove, CA: Brooks/Cole.

Hansen, J. C., Stevic, R. R., & Warner, R. W., Jr. (1982). *Counseling theory and process* (3rd ed.). Boston: Allyn and Bacon.

Ivey, A. E. (1988). *Intentional interviewing and counseling* (2nd ed.). Pacific Grove, CA: Brooks/Cole.

Martin, D. G. (1989). *Counseling and therapy skills.* Prospect Heights, IL: Waveland.

Patterson, C. H. (1974). *Relationship counseling and psychotherapy.* New York: Harper & Row.

Robinson, F. P. (1950). *Principles and procedures of student counseling.* New York: Harper.

Rogers, C. R. (1965). Client-centered therapy. In E. Shostrom (Ed.), *Three approaches to psychotherapy* (film). Santa Ana, CA: Psychological Films.

Shertzer, B., & Stone, S. C. (1980). *Fundamentals of counseling* (3rd ed.). Boston: Houghton Mifflin.

Sommers-Flanagan, J., & Sommers-Flanagan, R. (1993). *Foundations of therapeutic interviewing.* Boston: Allyn and Bacon.

Wallace, W. A. (1986). *Theories of counseling and psychotherapy.* Boston: Allyn and Bacon.

CHAPTER 7

COMMITMENT TO ACTION AND TERMINATION

The process of self-exploration and accurate diagnosis of the problems with an empathic and perceptive counselor is sometimes sufficient to allow a client to identify the necessary actions to solve the problem and implement real change. This pattern is most common when the difficulty is short-lived and circumscribed and the client has a strong set of coping skills and a good support system. At other times, insight is not sufficient for change. This tends to happen when the problem is more severe and personal resources are limited. In such a situation, the activities inherent in the second stage of the counseling process bring the client to the point of understanding self better but do not clearly identify a single set of actions to resolve the concern. The client is no longer so confused about the issues, has more insight into the sources and patterns of the problem, and has experienced some release from emotional tension. However, the goals for change are still rather general, and the course of action to achieve them is still rather vague. Or, the actions required for change are fairly clear but entail risks that seem threatening to the client.

This chapter focuses on the third stage of the counseling process—commitment to action and termination. It highlights the tasks of defining specific outcome goals for counseling and then using those goals to design action plans to achieve them. Also discussed are the counselor's and client's tasks while the interventions are being implemented, the need for evaluation of the effectiveness of the interventions, and the counselor's responsibilities when the action plans don't get implemented or don't work as intended. Finally, the process of ending counseling, usually referred to as termination, is described, and resistances to termination and options for ending the process on a positive note are examined.

THE PROCESS OF GOAL SETTING

The first task of the third stage of the counseling process is goal setting. At the end of the second stage, counselor and client should have a clear diagnosis of the difficulties that brought the client to counseling and a good sense of the client's strengths and resources for coping with the problems. For example, at the end of the second stage, a male client who entered counseling because of loneliness and unsatisfying personal relationships understands the way he may be distancing other people, the way he may misinterpret their desire for independence as rejection of him, and his tendency to think negative thoughts about himself and thereby, heighten his feelings of loneliness and frustration. It is also likely that he sees how this pattern developed from his early relationships and how he acts to maintain that interpersonal style in current relationships. However, exactly what to do about this pattern or what he really wants from his relationships is not entirely clear yet. When asked, the client may respond that his goals are to get closer to people and feel less lonely. It is the job of the counselor at this point to help the client translate these general goals into more specific goals. Does he want a long-term intimate relationship with one person? Does he want to have more friends? Does he want to make new friends more easily? Does he want to know how to respond better to stressors and conflict in relationships? Does he want to be able to share his emotions and personal beliefs honestly and ask for support when he needs it? Does he want to be more giving to others and less self-interested? How will he know that he's not lonely anymore; that is, what internal and external markers will signal change? If he wants many of these things, which takes priority in his view? As you can see, what first appears to be a clear goal may have a number of different meanings to the client.

The process of specifying goals ensures that both client and counselor know exactly where they are headed in the third stage—enabling them to choose appropriate intervention strategies. The task of specifying goals has been compared to identifying the route one is taking on a trip. A common expression is "If you don't know what road you're taking, you may end up somewhere else." The most satisfying trips are usually carefully planned beforehand. The pre-trip plan need not be rigidly adhered to, but a sense of the steps to the destination is not lost in the deviations from the plan. This is a fitting analogy for the task of counseling at the beginning of the third stage.

What are the signals that the timing is right for a transition from diagnosis and in-depth exploration to goal setting and commitment to action? The most important marker is that the six components of diagnosis have been attended to and the process of self-exploration is no longer yielding new insights for the client. There is also sometimes a change in the client's statements, indicating restlessness with the status quo and a need to change. In the case of a person who has entered counseling because of depressed feelings, the counselor may hear more comments that the client is "tired of feeling this way" and is prepared to do what it takes to change even if risks are involved. There may be less hopelessness and more frustration, even irritability with the current state of affairs. The emotional release of sharing the pain with the counselor has become less rewarding, and the need to focus on the future is stronger. Fear of change and resistance to change do not disappear, but a new sense that it is time to "get on with things" is added to those feelings.

At this point, the counselor can shift his or her attention from diagnosis to goal setting. When a client has come to counseling to make a decision about an educational path or a career choice, the goals are fairly explicit from the start, and the transition to action planning is typically smooth. However, when a client comes with emotional pain or problem behaviors, the initial motivation for counseling is to stop feeling so bad or acting in such unproductive ways, and the goals are often quite vague. For example, a married couple may enter counseling because they want to stop arguing so much. This goal is entirely sufficient for the early stages of counseling; however when exploration and diagnosis are complete, such a goal is unlikely to generate productive action plans. The couple needs more clarity about what kind of relationship they want instead and how much less arguing will be defined as success. Plainly put, the question is "What else do you want to think and feel about each other besides anger, and what else do you want to do besides argue?" The transition to specific new patterns of thinking, feeling, and acting needs to be guided by the counselor. In this situation, the counselor begins to use selective reflection, confrontation, and what Cormier and Hackney (1993) call "ability-potential responses" to specify the goals more clearly. Because the use of confrontation is described in detail in Chapter 4, the discussion here focuses on selective reflection and ability-potential responses.

Selective Reflection

When a counselor uses selective reflection, he or she chooses to respond more fully to the part of a client's statement that shows yearning for change. The focus is still on the affective message the client is sending, but the counselor responds to that message selectively. Here are some examples:

Interaction A

CLIENT: I really hate it when I yell so much at my daughter. I know it's not healthy for either of us and it doesn't match with the way I want to relate to her.

COUNSELOR: Sounds like you have a picture in your mind of how you would like to act with her.

Interaction B

CLIENT: I'm sick and tired of being alone in my room when I know other teenagers are out having fun.

COUNSELOR: Sounds like you are fed up with the old pattern and are ready to do more of the things others your age do.

Interaction C

CLIENT: One thing has become clearer to me over the last few weeks in rehab. No matter how unfair it is that I'm stuck in this wheelchair or how angry or sad it makes me, I'm still going to be in this damn chair for the rest of my life.

COUNSELOR: You're becoming more aware now of the reality of your disability, and you seem to be looking for something else to fill your life besides rage and grief.

In each of these examples, the counselor reflected back affective content to the client, but the counselor's statements were centered on that portion of the feeling associated with readiness to change. These are, in essence, advanced-empathy responses as discussed in Chapter 4. If you look at Interaction A closely, you will see that the counselor responded to only one part of the affect the client expressed; thus, the response is a selective reflection. An ordinary reflection would have attended to the anger and worry implicit in the statement as well. It is important to note that the counselor in this example is still reflecting the feelings expressed by the client. No feelings are being suggested that are not at least at the edge of the client's awareness. A counselor who tries to impose readiness-for-change feelings on a client is likely to fail.

Selective reflection is used effectively when the client has progressed to the point of insight into self; clients who are pushed in this direction by their counselors usually respond with resistance. When the client feels he or she is not freely choosing a goal or action plan, the part of the self that fears change gains strength, and the counseling progress stops until the client feels in control again.

Exercise

As an exercise, write selective reflections for the following client statements and share them with a partner for feedback.

1. I'm beginning to see that if I stay with my husband I'm likely to get hit again and again. He's not going to change. It's not so scary to accept that now.
2. Wow! I'm really enjoying this internship in biochemistry. It's so much more fun than the internship in electrical engineering I did two years ago. I can see this enjoyment of biochemistry lasting a long time, too. (Statement of a college senior in career counseling.)

3. After the last few counseling sessions, I've been able to catch myself responding to my partner in those passive-aggressive ways. The pattern is clearer and I'm tired of repeating it.

Ability-Potential Responses

Another way to clarify and specify goals is to use ability-potential responses. According to Cormier and Hackney (1993), an ability-potential response is "one in which the counselor suggests to the client that he or she has the ability or potential to engage in a specified form of activity" (p. 111). Ability-potential responses may be used when the goals have been decided but the action plans to reach those goals are still under consideration. Consider the following examples in which the counselor gives an ability-potential response:

Interaction A

CLIENT: I could approach my professor about an extension to complete the course after my husband is out of the hospital, but this professor doesn't seem very understanding about the lives of students outside the classroom.

COUNSELOR: That's one good alternative. I can give you information about the college's procedures for getting course extensions. Let's use the next few minutes to think of other alternatives and then you can choose the one you think best.

Interaction B

CLIENT: I'm sick of being alone. Next Friday I'm going to a bar to see if I can meet a woman there. I'm not likely to meet anyone if I stay in my apartment.

COUNSELOR: You're right. Putting yourself in a social situation where women are around is a first step to getting to know someone. Are there other social situations in addition to a bar that might also help you achieve this goal? Let's brainstorm about them and then you can make the decision about which alternative you prefer.

Interaction C

CLIENT: I want to let my friend know how much she has hurt my feelings, but I'm afraid I'll use all the wrong words and risk the whole friendship.

COUNSELOR: I've noticed that in counseling you have been able to let me know when I have frustrated you and that you've done it without blaming me. How can you use your success with direct communication with me in your conversation with your friend?

The purpose of each of these counselor responses was either to help the client become aware of his or her capabilities or to expand the possible goals and choices generated by the client. When an ability-potential response is used to expand the available choices in a given situation, the client also learns a more adequate form of problem solving. All too often, what causes people pain and dysfunction is not so much the problems in living that they encounter, but a constricted style of trying to solve those problems. The expectation is that the process of exploring a wide variety of alternatives and weighing the merits of each will help clients learn a new way of approaching difficulties that will generalize to other issues in their lives.

In summary, the diagnostic and exploration processes help the counselor and the client assemble all necessary information. Confrontations and selective reflections further mobilize the client's readiness for change, and ability-potential responses structure the goal setting so that all possible alternative solutions to the problem are weighed.

What is involved in the process of weighing alternatives? It involves discussion of what the client sees as the likely outcomes of the possible alternatives and an evaluation of the degree of difficulty and desirability of each outcome. Schuerger and Watterson (1977) present one model for assessing the utility of each of several courses of action. They say that two variables contribute to utility: (1) the probability that the course of action will result in success, and (2) the value that the client assigns to succeeding with that alternative. In the end, what the client chooses should have a high value to the client and a reasonably good chance of success.

Sometimes, ability-potential responses get confused with advice giving, a counselor response that is rarely useful. The underlying goal in giving advice is to get the other person to do what you think best. Advice giving often feels to the client as though the counselor is taking away some degree of autonomy in decision making. It is important to keep in mind that the client has probably received lots of unhelpful advice from others before he or she sought counseling. In contrast, with ability-potential responses, there is no intent to compromise the client's autonomy or take on a parental role; rather, the counselor's motive is to respect the free choices of the client and assist him or her in effective problem solving.

One other aspect of ability-potential responses deserves mention. Implicit in an ability-potential response is an affirmation of the client's readiness to change and capacity to find a workable solution. The counselor is making an indirect statement of belief in the client's ability to resolve the problem.

The affirming aspect of this type of counselor response is particularly apparent in Interaction B above.

Counselor Directiveness in Goal Setting

There is debate among counseling professionals about the degree of directiveness a counselor should show in goal setting and action planning. One part of the debate centers on whether a counselor ought to suggest goals and actions to a client. (No legitimate theory recommends giving advice in a paternalistic fashion.) On the one hand, the argument is that counselors have a responsibility to share their expertise with clients, and failing to offer goals or alternative choices seems to deny that expertise and be incongruent with an open and trusting relationship. On the other hand, some theorists argue that client goals or actions that come from others are not truly "owned" by the client and are likely to meet with resistance. The resistance is sometimes shown by direct and open rejection of the suggestion and at other times expressed indirectly through "Yes, but" comments about the suggestion. The following example illustrates a "Yes, but" response. A counselor suggests that a client in career counseling may benefit from interviewing some people working in the careers the client is considering. The client's immediate response is to generate excuses about why such interviewing can't happen this week, or next week, and so on. When the counselor pursues other alternatives, a pattern of client responses beginning with "Yes, but" occurs: "Yes, but that's my mother's birthday and I need to get the party ready that day," or "Yes, but I have a calculus test that week." In this situation, the client is feeling uncomfortable with the suggestion but is unable to express that discomfort directly. Indeed, sometimes clients who get into this pattern are unaware of their resistance to the counselor's proposal. When this happens, the counselor needs to redirect the discussion or gently confront the resistance.

The important factors in deciding whether to suggest a goal or alternative to a client are the depth of the trust in the relationship, the degree of the client's ambivalence about change, and the nature of the problem itself. In a trusting relationship with a client who is highly motivated to change, the presentation of a goal or alternative by the counselor has a good probability of being well received and fairly considered. For example, a client who has entered counseling devastated by his wife's breast cancer and feeling all alone and fearful may welcome a suggestion that he consider attending a cancer support group that includes family members. (Note how different this approach is from advising the client that he ought to attend such a support group.) In addition, when a client enters counseling with a skill or information deficit (as is often the case in educational or career counseling), the merit of suggesting a goal or alternative to a client is also strong since there is little likelihood that the client will independently generate such information. However, when trust is not deep, the fear of change is almost as strong as the desire for it, or the problem involves deep emotional distress or long-standing maladaptive thoughts and behaviors, the wiser course of action may be to elicit alternatives from the client. The case of the client who repeatedly offers "Yes, but" responses illustrates the problems with suggesting goals when a client has some ambivalence about change and decision making.

Sometimes a process of brainstorming about all the possible approaches to resolving the problem can be used effectively. In such a brainstorming session, the counselor and the client work together to generate a large pool of alternatives. After the brainstorming session, the task is evaluating the merit of each possibility. Here, too, there is no expectation that any alternative the counselor has mentioned will be imposed on the client. In any case, the goals or alternatives suggested by the counselor need to be framed in the context of the diagnostic information about unsuccessful attempts to solve the problem. Suggesting an alternative that appears similar to a failed solution is obviously not likely to be acted upon by the client.

Whether to suggest goals or action plans to a client is also partially dependent on the theoretical orientation of the counselor. Humanistic theories are reluctant to be so directive, but cognitive and behavioral approaches view such activity as completely appropriate. Chapter 12 illustrates these differences more fully. Clearly, what needs to be avoided is a pattern in which the clients end up feeling as though they are guessing at an approach or goal the counselor already has in mind. Such a way of interacting is artificial and unproductive from any theoretical perspective.

By including ability-potential responses, confrontations, and selective reflections in his or her repertoire of responses, the counselor can focus the client's attention on goals rather than problems and then, if necessary, divide the goals into achievable subgoals. The goals or subgoals then lead to action plans for change. Consider the following case.

Case: Dolores

Dolores was a 50-year-old widow when she came to a counselor to talk about her frustration with her 25-year-old son, who was still living at home. The son, the youngest of three children, had friends, a full-time job, a car, and other trappings that go with young adulthood. Dolores's problem was that her son would do nothing around the house and expected to be taken care of as he had been when a child. He left his dirty clothes on the floor in his bedroom, did not assist in meal preparation or any other household management tasks, and spent any time when he was at home working on his car or hanging around "underfoot." Dolores had repeatedly asked her son to take responsibility for his own things, to maintain his own room, and to help with other household chores. She had refused to provide maid service for him, but the mess that accumulated was intolerable to her, and so she cleaned it up. Basically, Dolores wanted out of the "child" care business and felt that she had done her duty for all three children. She also realized that this arrangement was not healthy for her son either. The obvious solution was to tell her son that he would have to move, but that would be stressful because "a mother doesn't kick her own son out of the house." She felt hopeless about change.

The process of in-depth exploration and diagnosis by Dolores and her counselor resulted in a definition of the problem as a fear of confronting her son about his unacceptable behavior, an exaggeration of the negative consequences if she took such an action, and a lack of practice in assertiveness with family members. She was in good health and had reasonably good coping skills and an adequate support network, in spite of the loss of her husband. Through in-depth exploration, she also came to see more clearly that she was doing her son no favor by letting him stay in a childlike role with her. She came to differentiate between her style of acting with family and her style with other people. Dolores saw that with strangers or acquaintances she had little difficulty expressing her preferences and not acting in an overly submissive way. This healthy adult-to-adult style of interacting had not transferred to her relations with her grown children or other relatives. In addition, in-depth exploration had also given her insight into the role her beliefs about mothering had played in her nonassertiveness. Dolores modified these rather rigid beliefs to see the difference between promoting her son's welfare and simply submitting to his wishes. Through selective reflection, confrontation, and ability-potential responses, Dolores and her counselor generated these goals:

- To express to her son the need for a change in their living arrangements and her desire to have him find an apartment within two months without acting angry or sad while expressing this desire. (Dolores also examined two other alternatives before deciding on this goal. The first was to continue with the current situation while modifying her attitude toward it, and the second was to try to get her son to change his behavior at home. The first was quickly rejected, but the second was seriously considered. In the end, though, Dolores concluded that it had a low probability of success and even higher stress than asking him to leave.)

- To use positive self-statements while preparing for this conversation with her son and subsequent to it to remind herself that her desires are legitimate and probably better for her son in the long run.
- To generalize this assertive behavior with her son to her style of interacting with her parents, her other grown children, and other relatives.

If Dolores's case had been more complicated, with a larger set of problems or a wider range of alternative solutions, the use of a chart of apparent alternative goals with a listing of the pros and cons of each might have been helpful. Counselor and client can work together in developing such a chart, or the chart can be assigned to the client to work on between counseling sessions. With this approach, the client learns a more constructive method of problem solving and gets the message that the counselor believes the client capable of change.

DESIGN AND IMPLEMENTATION OF ACTION PLANS

Once goals are agreed upon and the client has expressed a commitment to them, the next step in the counseling process is to decide on a set of action plans that will help achieve them. Just as the process of goal setting involves careful evaluation of alternative goals, the process of action planning includes a careful review of alternative actions to help the client reach a goal and then a mutual agreement to implement the chosen actions. An *action plan* is a specification of actions the client will take (with the help of the counselor) to reach a goal. These terms imply change in visible behavior, but they are also used to describe strategies designed to facilitate changes in thinking or feeling. Action plans are initiated in the counseling session, but a good portion of their implementation involves activities outside of counseling. After all, counseling amounts to no more than one or two hours each week. Real change requires more practice than is allowed within the sessions. In addition, the transfer of learning to other settings is critical if the counseling goals are to be achieved. Sometimes these structured, outside activities are called "homework."

In the case of Dolores, the action plans were (1) an assertive discussion with her son about her decision, (2) the development and use of positive self-statements when self-doubts were felt, and (3) the use of assertive statements with other relatives when appropriate. To implement these action plans, Dolores needed information about assertiveness, practice in acting assertively, a set of positive self-statements that she believed in and was prepared to use, and some guidelines for timing the use of the positive self-statements. The agenda of the third stage of counseling focused on these tasks. The counselor had choices about how he would help Dolores become more assertive. Dolores could join an assertiveness training class, read books on assertiveness or be guided by her counselor to learn assertive behavior. The counselor assessed which plan seemed to best fit Dolores's situation and then listened to her assessment of the best plan. Next, they worked to reach consensus.

A counselor has a responsibility to present all the available interventions to a client, explaining how each plan could be carried out, the possible merit and drawbacks of each, and the probable time involved in implementing each. If the counselor believes one plan is superior to the others, that judgment should be shared with the client. However, unless the client chooses a path the counselor sees as harmful or counterproductive, the counselor agrees to carry out the alternative selected by the client. Because Dolores seemed to demonstrate appropriate assertiveness with persons outside her family, an assertiveness training class seemed unnecessary. Similarly, since few books on assertiveness focus on family relations, this alternative did not appear to be either time- or cost-effective. The alternative of practicing assertive behavior with the counselor, who can role-play family members and anticipate especially challenging situations during role playing, appeared to have the most merit. Then a plan was set up for Dolores to practice assertive action with family members between sessions and report her success or difficulties to the counselor. Since the client concurred with this judgment, the action plan was implemented.

Action planning requires both judgment and skill on the part of the counselor. Judgment involves differentiating between workable and unworkable plans, and skill involves knowledge and experience in using designated methods to carry plans out. A counselor who has only one of these attributes is unlikely to be able to help the client make significant change. One of us remembers a powerful experience from her counseling practicum in graduate school. Supervision of her counseling was taking place through an audio connection in an adjoining room. The counselor spent a difficult session using a person-centered counseling approach to a client's career indecision. In this hour, the counselor showed unconditional positive regard, empathy, and respect for the client. Her reflections of feeling were right on target, it seemed. As she entered the "debriefing" with the supervisor after the session, the supervisor opened with this comment: "That was skillful Rogerian counseling with your client. Unfortunately, in my judgment, Rogerian counseling was not what that client needed at this time." The supervisor went on to explain his view and made a good case for an alternative approach to counseling. This example epitomizes the need for both skill in carrying out interventions and judgment in selecting appropriate interventions. At this stage in training, the counselor had plenty of the former, but not enough of the latter.

In the case of Dolores, the counselor would need to use professional judgment to determine whether the goals chosen by Dolores were realistic and likely to be helpful. Professional judgment would also come into play in deciding what kind of assertiveness-training method would be most suitable for Dolores. Finally, the counselor would need skill in conducting assertiveness training and in tailoring positive self-statements to Dolores's particular situation.

Beginning counselors are often overwhelmed by the variety of intervention strategies available to them. One helpful way to organize action planning is a template offered by Brammer, Shostrom, and Abrego (1989). They divide action planning into three categories: "(1) strategies for restructuring client self-perceptions; (2) strategies for reducing physiological and emotional distress; and (3) strategies for behavior change" (p. 172). The counselor's theoretical orientation will greatly influence the degree to which each of these three categories of interventions is relied on, but almost all counselors use interventions from each area to some degree.

Strategies for changing self-perceptions can be viewed as interventions that help modify a client's thinking about self. In the case of Dolores, the action plan that involved increasing the use of positive self-statements was a strategy to change her view of herself. Consciously telling herself accurate positive things would help to counteract the destructive and untrue negative messages that were part of her prior learning. In essence, one could view the whole in-depth exploration stage of the counseling process as a plan to change self-perceptions. What is distinctive about this intervention in the third stage of counseling is the explicit focus on changing self-perceptions as a goal of counseling. In the second stage, such changes often occur, but they tend to emerge from the exploration and diagnostic process rather than in response to explicit goal setting.

Strategies for reducing emotional distress would not be directly applicable to the case of Dolores unless she became highly fearful or angry about the confrontation with her son. Such strategies are applicable to assisting a client who suffers from panic attacks, for example. Relaxation techniques or biofeedback procedures fall into this category, as do strategies that involve expanding the client's support network or, in the case of severe anxieties or depression, using antianxiety or antidepressant medications.

Strategies for behavior change are apparent in the assertiveness-training plans Dolores and her counselor developed. She intended to change her verbal behavior by applying the assertiveness skills she used in other situations to her family interactions. In the case of the person who suffers from panic attacks, possible interventions include contracts with the client to gradually engage in feared activities, the initiation of pleasurable activities such as exercising to distract from the obsessive worry about whether a panic attack will occur, and the design of a plan for self-reward as each small change is accomplished.

The counselor and the client also need to decide whether involving other people in counseling would be a helpful plan. If Dolores's son agreed to come to counseling with her, would such an activity have merit? Would it be better than the other alternatives chosen? Still other intervention strategies could have been useful to Dolores. Try brainstorming with a partner about other intervention strategies that might have been useful with this client.

Evaluate Outcomes

In the final analysis, the quality of an action plan depends on the satisfaction that its implementation brings to the client. Only after the fact can the counselor know whether his or her hypotheses were sound and whether the decision-making process took into account all the important factors. In counseling about the management of interpersonal affairs, a counselor often has the opportunity to help a client evaluate choices by reviewing events that occur between counseling sessions as the client implements his or her decisions. In matters requiring long-term planning, implementation may occur over a period of years, and evaluation will be a prolonged process. A client who has experienced effective counseling is likely to have the resources to analyze progress and likely to seek the help of another counselor if new problems he or she faces are very complex.

To return again to the case of Dolores, she did ask her son to leave. He was not at all upset and in fact seemed almost to have been waiting to be "pushed out of the nest." He liked his new independence, and he checked from time to time to see how his mother was getting along. Dolores enjoyed the freedom of not struggling with her son about mundane things, but she was still not fully satisfied with her living situation.

Repeat the Process

If, after evaluating the outcomes of a choice or course of action, a client still experiences stress or dissatisfaction, the whole process must be repeated. It is possible that the lack of success resulted from oversights somewhere along the line. The problem may not have been defined properly, or the decision about which problem was of greatest importance may have been wrong. Perhaps there was insufficient attention to information about the client's preferences or skills. Some of the hypotheses about the likelihood of success or the value attached to certain successes may have been in error. The choice may have been made impulsively or implemented poorly. And maybe everything worked as predicted, but the solution of one problem produced another.

The case of Dolores provides a good illustration. Since she still seemed somewhat dissatisfied with her living situation, she returned to the counselor about six weeks after her son moved. A review of the decision-making process indicated that a good outcome had been achieved. Dolores had gained what she wanted and had not had to suffer the negative consequences that she had anticipated. The fact that her son so readily accepted the move and seemed happy with his new lifestyle affirmed Dolores's decision and alleviated any concerns she had about not being a good mother. In fact, she had more "good mother" feelings after having asked him to move than she had before.

However, Dolores had not anticipated that she would be lonely when her son left, even though this possibility was addressed in the original counseling. She now described her life as very dull. It no longer mattered to her whether or not the house was in order, because no one ever came to see it and she was so lonely that she didn't care how it looked herself. A whole new problem existed, and a new set of goals and action steps had to be considered. In a sense, one could say that a bit of information was overlooked in making the original decision, in that she did not know she would feel so lonely and sad. On the other hand, the favorable consequences of the decision affirmed its basic soundness. (In spite of her current difficulties, Dolores never regretted her son's departure.) Final resolution of the new problem came for Dolores when she decided to take in a housemate to share her home and add variety to her life. In the process of making this decision, she considered the prospects of moving somewhere where there were some group activities, living with other relatives, and other options. If she had been involved in a romantic relationship that had the possibility of leading to a long-term commitment, this factor would have been weighed into her goal setting. In the end, she decided that her new goal was to find a housemate because this action plan called for the fewest changes in her life—retaining the things she valued and providing company. The action plan involved helping her structure her search for a housemate so that it would maximize her chances of finding someone with whom she felt compatible.

Obstacles to Implementing Action Plans

It would be easy to conclude from the preceding discussion that clients who are helped through the systematic problem-solving approach move easily from making plans within counseling sessions to actions in real life outside counseling. Unfortunately, the change process is not that easy for many clients. There are two major reasons why people have trouble implementing behavior that they have helped to decide on: (1) any new course of action feels unfamiliar and includes the risk of failure, no matter how logically it has been derived; and (2) old familiar behaviors frequently have rewards as well as difficulties. Even though change may reduce the adversity, it may also reduce the rewards. These rewards are also referred to as "secondary gains."

The case of Dolores again serves to illustrate each of these points. First, there was the prospect of failure when she confronted her son with the decision that he must leave. She did not know if he would become angry and unreasonable. She was not even sure that he would leave if she asked him to do so. If he refused to leave and she could not convince him to do so, there was the potential for the situation to become even more intolerable. The potential for an intervention to backfire also presents an obstacle to implementing action plans. No matter how carefully thought out an intervention is, unintended consequences may ensue. In this respect, clients' caution about change is fully appropriate.

Second, Dolores had adapted her lifestyle to accommodate the behavior of her son. Even though she objected at one level to serving as caretaker to an adult son, her life had a predictable quality. She had a daily routine that obviously would change markedly if she asked her son to leave. This predictability was a secondary gain from the status quo. Another major deterrent to action for Dolores was the secondary gain she experienced from being able to think of herself as a "good mother." A good mother takes care of her "child" and gains a measure of self-identity through the process of being a good caregiver. A final reward of the status quo that did not become obvious to Dolores or her counselor until after it was removed was the fact that she enjoyed having another person in the house for companionship. Failure to understand that secondary gain led to the need for further counseling and probably some delay in establishing a satisfactory long-term living situation.

This analysis can be extended to Dolores's decision to seek a housemate. This decision brought up the question of how well she would get along with a stranger living in her house. The housemate could have personality attributes or personal habits as objectionable as those of Dolores's son. Furthermore, the rewards of living alone included privacy and the freedom to do whatever she wanted whenever she wanted to do it. These secondary gains would be sacrificed if someone came to live in her home.

Supporting a Client's Action Plan

Because it is hard for many clients to change behaviors in ways they think will improve their lives, counselors frequently find that they must lend support to a client's decision to act. This is initially done within the counseling session when the goal is translated into an action plan, but often it must be repeated in subsequent sessions.

Support for a client's action plan can take several forms. The counselor may dwell on the positive benefits the client will derive from acting and achieving desired outcomes, feeling in control of his or her own life, or eliminating unwanted hassles. The counselor may also work on reducing the client's fear of acting by reviewing the potential negative outcomes and helping the client to see that such outcomes might not be so difficult to handle. Clearly, ability-potential responses, selective reflections, and confrontations also serve to bolster the client's energy for implementing the action plan.

The counselor might ask the client to actually picture himself or herself employing some new behavior and describing the scene. By examining such imagery (mental pictures), both client and counselor can gain insight into the client's needs, aspirations, and fears concerning the specific situation. Picturing anticipated outcomes of a new behavior provides an opportunity to rehearse the behavior itself and ways to handle the responses of significant others. The counselor may choose to bring some scenes to life through role playing and give the client practice in dealing with anticipated reactions of significant others.

Support-for-action principles can be applied to the case of Dolores. The counselor validated Dolores's wish to be free of unwanted maid duties. He also emphasized her right to assert control over her own life and to establish a desirable lifestyle in her own home. Dolores was asked to picture the worst possible outcome of asking her son to move out. She saw him stalking angrily out of the house and returning the next day for his belongings. After some thought, she concluded that he would not become permanently estranged and that, undesirable as the scene was, she could stand his temporary hostility and absence if she had to. She thought it very unlikely that her son would refuse to move, and she felt that other people, especially her son's friends, could be called upon to reestablish contact if need be. With clients like Dolores, who have a history of coping effectively with life's difficulties, minimal levels of honest encouragement are usually sufficient to produce action.

With other clients, who find the solution to a problem to be rather painful or who have a history of being ineffective under stress, the process of mobilizing energy for constructive action may take longer. Frequently, such a client will leave a counseling session having decided on a course of action only to return to the next session having failed to act. In some cases, feeling embarrassed about failing to act, the client will interrupt counseling rather than face the counselor again. The possibility of such a negative outcome reminds counselors how important it is to design action plans carefully and to be alert to signs that the client may not be ready to implement them.

In still other cases, a client may have acted only to find that stress was still at a high level and the problem had become worse. Consider what might have happened in Dolores's case if her son had refused to move and had increased his demands for caretaking. Such behavior would not have been too unusual; a client's family will often react to changes in his or her behavior by trying to maintain the status quo. Dolores's son could have responded to her request for freedom with an insistence that she maintain her present role. If that had occurred, different strategies and substantially more encouragement would have been needed.

When a client returns to counseling (or is sought out by the counselor) after having failed to carry out the action plan, a reexamination of the plan is in order. The counselor might ask the client to consider what it was about the plan that made it seem so hard to carry out. Sometimes the need for small changes in a goal or plan are indicated and at other times a completely different, more feasible plan may have to be devised. When a client has difficulty coping with stress or is facing a complex situation that may be difficult to change, the counselor needs to help design action plans that are "successive approximations" to the goal. In other words, together with the client, the counselor must design a series of actions, each of which brings the client one step closer to the goal. For example, if a client's goal is to express anger more appropriately when in conflict with her mother, a successive approximation to the goal is to use the skills learned in counseling with siblings or friends first. The client is more likely to achieve success in such situations and more likely to handle small failures with these people. The successes and difficulties can be examined in subsequent sessions and the client's conflict management skills can be reinforced or modified. When confidence and skills are built, the client will have maximized her chances of handling conflict with her mother appropriately.

Direct Intervention

If a client's stress levels are very high and careful implementation of disclosure, exploration, goal setting, and problem solving fails to reduce the stress or produce any promising plan of action for the client, a counselor may decide to take steps to change some condition of the client's life.

There are a variety of possible situations in which a counselor may conclude that a client's present resources are inadequate to his or her circumstances and in which intervention will require manipulation of the environment. It is a heavy responsibility for a counselor to reach such a conclusion because direct intervention is likely to create a feeling in the client that he or she is in need of being taken care of and that other people will always step in to "fix" things when they go wrong. In short, a counselor's failure to help a client exhaust all of his or her own resources can cause the client to become dependent on others and fail to try to cope when the going gets rough.

Nevertheless, there are times when a counselor must step in to avoid unnecessary harm to a client. Situations in which professionals or caregivers act destructively represent one such circumstance and are more common than most people suspect. There are also instances when very depressed clients are self-destructive in their behavior and unable to judge for themselves what is best for them. Other clients may be so angry and out of control that the safety of other people is at stake unless the client is placed in a safe environment, such as a hospital. These examples serve to illustrate the kinds of "dead-end" situations in which a counselor must choose to act in behalf of the client. Unfortunately, aside from the situation in which the client or other people are in imminent danger, there are no easy rules as to when such intervention is required. The counselor must judge that the client is incapable of effective action and that failure to make some kind of change will result in significant harm to the client or other people.

It is common for a counselor to enlist the help of parents, spouse, teachers, and other significant people in the client's life to assist and support the client. Of course, unless there is real danger, this involvement takes place only with the consent of the client. Such environmental manipulation capitalizes on the caring and competence of these support people. The usual role of the significant others is to contribute to the reinforcement of the change process by carrying out plans that are developed by the client in consultation with the counselor and the significant others. This approach is used very frequently with children and is a mainstay of the work of the elementary school counselor (see Chapter 8). Nevertheless, it is important to involve the client in the planning and to give him or her as much responsibility for change as possible. In this way, the client develops coping skills and confidence in his or her ability that cannot be developed by having things done by others.

The Organizational Context of Goal Setting and Action Planning

The model we have presented thus far assumes that the decision about the course of counseling is in the hands of the persons involved. Unfortunately, such control and freedom are not always available. Counselors in community agencies operate under reimbursement constraints that often shorten both the diagnostic and intervention stages of counseling. Similarly, counselors in educational settings are often responsible for large numbers of students as well as having other job responsibilities, and so the time for counseling is quite limited. Even counselors in university settings, who are often free of concerns about insurance or other administrative tasks, experience budget pressures that can limit the number of sessions with any individual client.

In the long run, counselors need to get politically active to modify policies that restrict individual access to counseling. In the meantime, counselors need to be clear in communicating to their clients about the constraints that are operating, to focus their attention on the issues of most concern to the client, and work to hone their counseling skills so that the time that is available is used with maximum effectiveness.

TERMINATION

Ending counseling on a positive note is the final task of the third stage of the counseling process. Ideally, termination occurs by mutual agreement when the goals set out by the client are achieved to the satisfaction of counselor and client. The client has grown in the ways he or she wanted, and often in unexpected yet desirable ways. In addition, in the process of solving the current problem, the client has deepened his or her self-understanding and broadened his or her coping skills; thus, other difficulties that arise will not seem so overwhelming. In other words, the client is better able to transfer learning from this situation to other problems.

There is another aspect of termination that is important for counselors to recognize. Termination also means achieving closure on the counseling relationship (Ward, 1984) and dealing with the inevitable issues of loss when that relationship has been intense or prolonged. Readiness for termination (which will be discussed in more detail later in this section) is a function of these relationship issues almost as much as it is a function of the achievement of outcome goals.

We recognize that termination may also occur because of an independent decision by the client, because of unavailability of the counselor, or because of institutional factors unrelated to either party, such as the end of insurance reimbursement or the end of the school year. There are also times when counseling needs to be interrupted because of an illness or long absence on the part of the counselor. [An article by Seligman (1984) deals with this kind of temporary termination and is a helpful resource in such situations.] In these situations, the principles we describe below do apply, although implementing them often requires ingenuity and persistence. Because clients often drop out of counseling or are forced to go through termination by outside factors, the counselor ought to be continually aware of the possibility of a premature end to the relationship and intermittently focus with the client on what has been accomplished thus far in counseling. When premature termination is a possibility, counselors and clients who think in terms of short-range and long-range goals will be more likely to feel some satisfaction and closure when counseling ends, even if the termination date is artificially imposed.

Readiness for Termination

It is time to terminate counseling when the client has achieved what he or she wants from the experience. Signs of readiness include positive and identifiable changes in the client's behavior, positive and pervasive changes in the client's mood, consistent reports of improved ability to cope with stress, and clear expressions of commitment to verbalized plans for the future. Important but less obvious signs include a sense of relief and an increase in energy. Generally, the client is ready for termination when he or she consistently responds in desirable ways in a variety of situations, including reasonable challenges. These signs of readiness for termination begin to appear before the final session, and counselors need to prepare clients for termination before the last interview. How much preparation is required is largely dependent on the intensity and length of the counseling relationship. For long-term clients, termination should be discussed weeks in advance of the last session. For example, Shulman (1984) recommends that termination from long-term counseling should constitute about one-sixth of the counseling process, and Ward (1984) suggests that termination from counseling takes on special importance when the issues the client brought to counseling deal with dependency and separation from significant others.

With many clients, readiness for termination is not difficult to assess. The client reports changes that have occurred, indicates a clear sense of having acquired what was wanted from the experience, and shows clear signs of being ready to end counseling. The changes reported by the client are consistent with the counselor's observations. At this point, the idea of termination may be brought up by either counselor or client. In circumstances such as these, the final counseling session is usually devoted to a review of the major themes, changes, and critical moments that have occurred during counseling. One of the counselor's goals in termination is to help clients gain deeper insight into just how far down the road of change they have traveled. Honest expressions of support are offered by the counselor to encourage the client to maintain the changes he or she has implemented. Depending on the circumstances, the counselor may recommend a follow-up session to assess whether or not the client has been able to implement long-range goals and maintain behavior change with reasonable effectiveness. It is appropriate for the counselor to inform the client that he or she will be available in the future, and if the relationship warrants, the counselor may request feedback about the counseling process from the client.

With other clients, readiness for termination may not be so easy to determine. The client wanting to overcome depression may be depressed less often but may still get mildly depressed occasionally. The client wanting to behave more assertively may be able to respond assertively to some situations but may still have difficulty under more challenging conditions. The client who has clarified his choices about a career may be procrastinating about making final commitments. Under such conditions, the general principle is that if a client shows strong signs of insecurity about being able to maintain desired changes, he or she is probably not ready to terminate counseling. These conditions may suggest that other themes need exploring as part of second-stage work or that more practice with interventions is needed as part of third-stage work. Many counselors move to a less frequent schedule of counseling sessions to help the client become more self-reliant and self-confident.

When real change has been accomplished but a client expresses feelings of insecurity in the final session, the insecurity may be more related to difficulty handling the loss of the counseling relationship than anything else. Counselor and client need to work together to sort out the source of this insecurity. Sometimes, it is simply an indirect expression of the loss of the regular contact with the counselor. When this pattern emerges, it is important to frame the end of counseling as a step forward in client growth rather than a trauma (Pate, 1982, p. 188). Scheduling a follow-up session some time in the future may relieve some of the client's fears about handling the world independently.

When a counselor has a cognitive or behavioral theoretical orientation, the time for termination may be apparent from the beginning of counseling. Many counselors using these frameworks contract with clients for a specific number of counseling sessions. Contracts, of course, are always open to renegotiation as new information emerges, but the time-limited nature of the counseling relationship is apparent from the start in these cases. Those who operate from a humanistic or psychoanalytic tradition typically leave the process more open-ended. Thus, no specific ending date is obvious from the start of counseling.

Resistance to Separation

Even when it is clear that the client has acquired what he or she wanted from counseling, termination can sometimes be a difficult experience. This can be especially true when the counseling has extended over a long period of time and a high level of intimacy has been established. For both participants, the experience of separation may be a reminder of past experiences with loss, isolation, rejection, loneliness, fear of the loss of important gratification, and fear of self-reliance (Cormier & Hackney, 1993; Goodyear, 1981; Weinberg, 1984). Resistance may therefore come from both participants.

The Client's Resistance. To the client, the counselor may have become an anchor and a source of security in a life of stress. The experience of being cared for and prized by the counselor, the feelings of relief and restoration of hope, and the discovery of new sources of personal strength and new capacities may all create strong attachment bonds (Lowenstein, 1979; Weiss, 1973). Letting go under these conditions can cause the client to suffer an intense loss. The stress can be especially problematic for lonely clients who have not experienced much intimacy in their lives, for clients who have in the past lost hope or severely doubted their capacity to function, and for clients who have been helped to deal with especially difficult life situations.

Even clients who have made great progress in counseling may feel insecure about going on without the support of the counselor and may try to prolong counseling beyond the time when termination could reasonably occur. On occasion, clients revert back to problematic patterns in an attempt to maintain the counseling relationship. At other times, clients express their resistance to termination through anger at the counselor. Another way of showing resistance is to develop entirely new counseling goals. For example, a client who has been working on self-esteem and social skills may suddenly express a need for career counseling. When any of these "symptoms" of resistance appear, counselors need to get clients to focus on the source of these feelings and behaviors. Then, they can decide together whether the concerns are legitimate reasons for extending counseling beyond the agreed-upon termination date.

In order to prepare clients for termination, counselors ought to raise the topic well in advance of the end of counseling and not encourage too great a dependency on the counselor. Such preparations will help minimize resistance to termination. Counselors should be especially cognizant of their role as anchor and work to help clients establish close friendships and support systems before counseling is ended, as these will ease the depth of the loss. A counselor who has not helped a lonely and isolated client develop friendships has not really attended to a fundamental counseling issue for that client.

The Counselor's Resistance. According to Nystul, "the ultimate goal in counseling is for counselors to become obsolete and unnecessary to their clients" (1993, p. 36). Counselors who get rewarded for helping others and who come to cherish their clients may forget this basic concept. Thus, resistance to separation may also come from the counselor. Ward (1984) suggested that the relative scarcity of published material on the counselor's resistance to termination reflects the profession's discomfort with open discussion of this topic. He argued that ending is often a difficult issue for counselors to talk about. The writings of Kovacs (1965), Mueller and Kell (1972), and White (1973) suggested that people who choose counseling as a career may be especially high in needs for intimacy, for giving nurturance, for gaining acceptance, and for receiving acknowledgment of their competence. Keeping a counseling relationship going provides an opportunity for these needs to be satisfied. Ending a counseling relationship risks the loss of satisfaction. The very needs that motivate people to become counselors may create barriers when it is time to let go. Goodyear (1981) suggested eight conditions under which letting go may be especially hard for the counselor:

1. When termination signals the end of the significant relationship
2. When termination arouses the counselor's anxieties about the client's ability to function independently

3. When termination arouses guilt in the counselor about not having been more effective with the client
4. When the counselor's professional self-concept is threatened by a client who leaves abruptly and angrily
5. When termination signals the end of a learning experience for the counselor
6. When termination signals the end of a particularly exciting experience of living vicariously through the adventures of a client
7. When termination becomes a symbolic recapitulation of other farewells in the counselor's life
8. When termination arouses in the counselor conflicts about his or her own individuation (p. 348)

In his work on mourning and melancholia, Freud (1938) identified symbolic fear of loss as a significant unconscious source of resistance to letting go. He reasoned that during the earliest stage of human development (the oral stage), the infant's most basic needs are to take in food and to receive the emotional nourishment that accompanies it. If these life-sustaining needs are not met on a stable, reliable basis, fear of loss is established. As the child develops, this fear may be repressed but still active. In later years, real as well as anticipated losses may symbolize the conditions of original learning. Psychoanalytic counselors argue that termination of counseling can symbolize early loss for both the counselor and the client. It is also possible that the positive counseling relationship that is about to end has been a particularly important source of gratification for the counselor during a period of stress and pain in his or her own life. Under such circumstances, relinquishing a rewarding experience can be difficult regardless of early life experiences.

It is not appropriate for a counselor to put off termination in order to continue having his or her needs met. Such a stance violates the basic ethical principles of the profession—to do good and avoid harm. Counselors who repeatedly experience intense feelings of loss when terminating counseling or who foster dependency rather than independence in clients should seek counseling to clarify and resolve their resistance to separation.

Ending in a Positive Way

As mentioned above, in intensive and long-range counseling, preparation for termination begins well before the last counseling session. As the signs of a client's change become more evident, the issue of how many more sessions are needed emerges quite naturally. In this context, it is appropriate for the counselor to share his or her honest assessment of the client's growth. As the signs of completion become apparent, it is generally appropriate to schedule one or two more sessions. These sessions can be devoted to follow-up work and to helping the client apply what has been learned. Then, when the last session occurs, both client and counselor will be prepared. The expectation of ending will make termination easier and will probably lead to good closure.

The principles we have developed in this chapter lead to some important guidelines for ending an intense counseling experience in a positive way:

1. Be clearly aware of the client's needs and wants. Ending counseling is likely to have important effects on the client. He or she may need to talk about what it will mean and to work through feelings of dependency and the need for support.
2. Be clearly aware of your own needs and wants. You, the counselor, need to end the experience for yourself as well as your client. Some of your own needs may make you want to resist termination. Awareness of these needs can help you control this resistance. Be particularly alert for inappropriate intrusion of your own needs if you are experiencing a great deal of stress or loss in your life.
3. Be aware of your previous experiences with separation and your inner reactions to these experiences. If the counseling experience has been intense, you can expect your reactions to a separation to be similar to your reactions to previous separations. Expecting and being aware of these feelings can help you deal with them better.
4. Invite the client to share how he or she feels about ending the experience. Putting closure on the experience is important for the client. Sharing inner experiences about termination is an essential part of the closure process.
5. Consider sharing honestly with the client how you feel about the counseling experience. If counseling has been intense, there will have been moments of excitement, anxiety, confusion, and stress for you. Reviewing these experiences may help both you and the client with the closure work. Both of you should have learned from the experience. Not all clients can handle such sharing, but judicious use of it can be an affirming experience for the client.
6. Review the major events of the counseling experience and bring the review into the present. Reviewing with the client the major themes of counseling, changes that have taken place, and critical moments brings the experience full circle. This focus helps the client affirm that growth and change are part of life and gain greater perspective on

his or her changes. Seeing self over time helps create closure. Using "I" messages ("I remember when we first started" or "Some moments that seemed especially important to me are . . .") helps to personalize the summing up. (A mini-review of progress is useful intermittently during the counseling process, not just at the ending session.)

7. Supportively acknowledge the changes the client has made. Implementing changes and using new behaviors is never easy, natural, or automatic. Maintaining changes requires affirmation and encouragement. Letting a client know that you recognize the changes he or she has implemented and the stress involved in doing so encourages the client to maintain these changes after the completion of counseling. When a client has chosen not to implement action plans for other issues that emerged during the in-depth exploration phase of counseling, the process of termination should also include an inventory of such issues and a discussion of the option of future counseling should the client see a need to implement changes in those other domains.
8. Request follow-up contact. Such a request not only provides you with information about the long-term effectiveness of the counseling strategies you used, but also demonstrates a sense of respect and caring for the client. Follow-up contact can be accomplished in person, over the telephone, or through correspondence. Personal contact obviously provides a stronger message of caring, but any of the methods gives information about counseling effects. The knowledge that follow-up will occur also acts as an additional incentive for the client to maintain the changes that counseling has produced. The timing of follow-up varies with the client's circumstances, but three to six months after termination is typical. A second follow-up may be conducted a year after termination.

Referral

Counseling is not always a successful experience. Client and counselor may have incompatible personalities (Goldstein, 1971), the client's difficulties may be beyond the counselor's helping skills, or the client's difficulties may require special modes of intervention (e.g., in cases of incest or chemical dependency). Under such conditions, referral is an appropriate choice for the counselor. On the part of the counselor, referral requires both an honest acknowledgment that some other person or resource in the community may be more helpful to the client and a willingness to help the client make contact with these resources. Effective referral requires that the counselor have accurate information about resources in the community, including knowledge about the scope and quality of their services. Since it can be very hard to tell one's story to a stranger whose qualifications are unknown, having personal contacts and the names of specific people in an agency can make the transition easier for the client.

Helping clients work through resistance to seeing a different counselor or other professional is often an important task in the referral process. Most people do not want to hear that they need special help; even if there is agreement with the idea of special treatment, they may still resist. The referring counselor should expect to spend at least one full session helping the client work through resistances. Helpful counseling techniques include support of the client, gradual presentation of the idea of referral, empathic acknowledgment of themes that emerge in response to the idea, a willingness to share firsthand information about the resource involved, and a willingness to either make an appointment for or go with the client. (Of course, no information can be shared with the referral person until the client gives informed consent.) Depending on the level of trust in the client-counselor relationship, it may be appropriate for the counselor to ask the client for follow-up feedback. This offers support to the client and also gives the counselor information on which to base a decision about making future referrals to the source involved. In case the match with the person to whom the client is referred turns out not to be a good one in the client's eyes, the client should know that the counselor is available to give additional referral sources. Giving a client several referral sources from which to choose is also a prudent move, since it reaffirms the client's autonomy and compensates for any unexpected difficulty in getting an appointment with the referral person.

When a client is not making progress, the counselor must determine whether termination or referral is appropriate. In instances in which the client is experiencing significant stress, referral should be attempted. Continuing to work with a client when that person is not being helped by counseling is unethical. Similarly, failing to provide for an appropriate referral in such cases is also unethical and can be seen as abandoning a client.

SUMMARY

The counselor promotes commitment to action in the first part of the third stage of the counseling process. Sometimes the sharing of concerns during the disclosure and exploration stages of counseling will have reduced tension and suggested obvious goals and actions. In such cases, the third stage may be brief or nonexistent. Frequently, though, the client does not spontaneously solve his or her problems during the earlier stages of counseling, and specific goals and action plans must be developed and reinforced during the final stage. Three types of counselor responses are especially helpful in defining specific goals and action plans. The first is selective reflection, attending to the affective message that suggests a client's readiness to change. The second is confrontation, pointing out the discrepancies between current circumstances and the client's goals. The third is the ability-potential response, a counselor response that helps the client generate alternatives to current behavior and offers reinforcement for readiness to initiate real change.

Clients are fearful of change, even desired change, and they usually require encouragement to act on their plans. Ability-potential responses offer this encouragement. The counselor supports the decision to act and helps the client to rehearse new behaviors. Sometimes the client may be in a position in which no solution appears to exist, or it may appear that the client can do some things for himself or herself but would make more rapid progress if significant others assisted. In such cases, the counselor may manipulate the client's environment by securing the intervention of other people.

Some courses of action bring relatively quick success. In other cases, goals and action plans must be reviewed and refined. In still other cases, the results of action plans will not be known until the client has lived with the decision for a period of years. Nevertheless, changes in a client's thinking, feeling, and behavior are important goals of counseling, and the effectiveness of the client in mobilizing resources to accomplish his or her goals is a tangible measure of success for both client and counselor.

Termination comprises the second part of the third stage of the counseling process, and it is a critical point in that process. Effective termination provides a positive closure to the experience and encourages the client to continue using his or her new learning. Because termination is an experience of separation after an intense relationship, feelings of loss can sometimes create resistance. The counselor should try to be aware of his or her emotions as well as those of the client and to communicate honestly about present emotions. Offering a personalized review of the overall counseling experience is often helpful. Guidelines for helping the client work through resistance to referral include giving support, presenting the idea at the proper time, and having reliable information about the resource involved.

If counseling is not successful, referral may be an appropriate course of action. Professional ethics demand that all reasonable action be taken to be sure that the client gets needed services.

REFERENCES

Brammer, L. M., Shostrom, E. L, & Abrego, P. J. (1989). *Therapeutic psychology: Fundamentals of counseling and psychotherapy* (5th ed.). Englewood Cliffs, NJ: Prentice-Hall.

Cormier, L. S., & Hackney, H. (1993). *The professional counselor* (2nd ed.). Boston: Allyn and Bacon.
Freud, S. (1938). The psychopathology of everyday life. In *The basic writings of Sigmund Freud.* New York: Random House.
Goldstein, A. P. (1971). *Psychotherapeutic attraction.* New York: Pergamon.
Goodyear, R. J. (1981). Termination as a loss experience for the counselor. *Personnel and Guidance Journal, 59,* 347–350.
Kovacs, A. L. (1965). The intimate relationship: A therapeutic paradox. *Psychotherapy: Theory, Research and Practice, 2,* 97–103.
Loewenstein, S. F. (1979). Helping family members cope with divorce. In S. Eisenberg & L. E. Patterson (Eds.), *Helping clients with special concerns.* Boston: Houghton Mifflin.
Mueller, W. J., & Kell, B. J. (1972). *Coping with conflict: Supervising counselors and psychotherapists.* Englewood Cliffs, NJ: Prentice-Hall.
Nystul, M. S. (1993). *The art and science of counseling and psychotherapy.* New York: Macmillan.
Pate, R. H. (1982). Termination: End or beginning? In W. H. VanHoose & M. R. Worth (Eds.), *Counseling adults: A developmental approach.* Pacific Grove, CA: Brooks/Cole.
Schuerger, J., & Watterson, D. (1977). *Using tests and other information in counseling.* Champaign, IL: Institute for Personality and Ability Testing.
Seligman, L. (1984). Temporary termination. *Journal of Counseling and Development, 63,* 43–44.
Shulman, L. (1984). *The skills of helping* (2nd ed). Itasca, IL: Peacock.
Ward, D. W. (1984). Termination of individual counseling: Concepts and strategies. *Journal of Counseling and Development, 63,* 21–26.
Weinberg, G. (1984). *The heart of psychotherapy.* New York: St. Martin's.
Weiss, R. S. (Ed.). (1973). *Loneliness.* Cambridge, MA: M.I.T. Press.
White, R. W. (1973). The concept of healthy personality: What do we really mean? *The Counseling Psychologist, 4,* 3–12.

Part Two

Adapting the Counseling Process to Diverse Clients

CHAPTER 8

WORKING WITH CHILDREN AND THEIR PARENTS

In the broadest sense, counseling with children involves the same process as counseling with adolescents or adults: a caring relationship with genuine empathic understanding promotes disclosure; deeper-level empathy and confrontation with self builds insight; and action planning leads to better coping strategies. Nevertheless, the counselor who has worked only with adolescents and adults is likely to find that initial encounters with children in counseling require skills and understandings he or she may not possess. Most books on the counseling process, and until recently most counselor education programs, have ignored the special qualities that children bring to counseling.

This chapter presents an introduction to counseling children with the clear recognition that those seeking to specialize in working with children need to extend their study well beyond the scope found here. The purpose of this chapter is to apply the general principles of counseling to work with children and to suggest some of the issues that form the fabric of a specialization in child counseling. Since all counselors work with children or with the parents and older siblings of young children, we believe that all counselors need a basic understanding of the helping process as it applies to young children.

For purposes of definition, we will refer to children as persons who have not yet reached puberty. However, it is obvious that there are big differences between 4-year-olds and 10-year-olds. As a general principle, the more mature the client, the less it is necessary to employ the special counseling procedures described in this chapter and the more applicable are the standard procedures described in earlier chapters of this book.

HOW CHILDREN DIFFER FROM ADULTS

The first quality of children that the counselor will notice is a limitation in verbal skills. The young child has a vocabulary that allows for labeling of persons and objects and for describing simple happenings. But according to Nelson (1966), "In contrast with his older siblings who can *verbalize* frustrations, love, anger, and acceptance, the younger child *acts* his feelings. He crashes cars together, he hugs his Mom, he shoots the enemy, and he hands another child a toy (p. 24)." A child may not be able to verbalize his or her emotions. Any statement of a problem is likely to be stated in simple terms with little clue to the cause or surrounding circumstances. In fact, children who are brought to counseling (in contrast to mature clients) "may not agree or recognize that there *are* problems or concerns" (Prout, 1989, p. 16).

Verbal skill is, of course intimately related to cognitive functioning, and if the counselor is able to sustain counseling with a young client long enough to encourage disclosure, immature cognitive processes will become apparent. "Children in their first few years of life think mainly in concrete terms and have not yet acquired the capacity for abstract thought" (Barker, 1990, p. 4). According to Piaget and Inhelder (1969), children do not become capable of problem solving through hypothesis testing until age 11 or later. Prior to that time, the child develops through several more basic stages that are characterized by initial acquisition of language, egocentrism, and problem solving focused on tangible objects. The understanding of problems of human interaction that have clear cause-effect components for the counselor may be beyond the comprehension of the child.

Two elements of cognitive development that often prove confounding to the efforts of an adolescent- or adult-oriented counselor who undertakes work with a child are the limited time perspective and the limited ways of viewing right and wrong. To realize the time perspective problem, one need only remember the last time one took a long automobile trip with a child or promised that his or her birthday party would be in one week. In both instances, the child probably began inquiring almost immediately about the culmination ("When will we be there?" or "Is my birthday going to be here tomorrow?"). No amount of explaining will convey accurately that "We will be at Grandmother's in three hours" or that "The sun must come up seven more times before your birthday." To a child, the world is eternal, and he or she is at the center of that world. To the counselor, time is finite, and one important responsibility is helping clients think about and plan for the future—a concept that is beyond the capacity of young children to understand.

The child's early conception of right and wrong, according to Kohlberg (1964, 1981) is focused on the avoidance of punishment and the satisfaction of immediately felt personal needs. The assumption (possibly correct for very young children) is that "others are in control, and I am not really responsible." At this stage of development, it is not yet really possible for the child to operate from an internal locus of control based on personal values or standards, because these have not yet developed. The young child will comply with demands from parents and teachers, or he or she will find ways to manipulate them to satisfy egocentric desires. Over time, children's moral decision making shifts to concern for how others will view them and how they view themselves (as good or bad children). By the end of childhood, the typical level of moral reasoning results in controls on behavior in order to abide by rules that contribute to the social good. (There are three further levels of moral development in Kohlberg's system that describe adults as exercising increasingly independent personal valuing in making decisions, though many adults never progress beyond the "law and order" stage.) Gilligan (1982) stated that females are more concerned than males with how their behavior affects their relationships with others.

The counselor working with children needs to think deeply about how a newborn child begins with virtually no understandings about relating to others and no skills and grows through stages to greater and greater levels of competence. Erikson (1963) described eight stages of human psychosocial development, four of which occur during the childhood years. First, children need to develop trust in their environment and caregivers. Then, they begin to develop some self-control and control over their environment, followed by a sense of initiative. During the elementary school years, their attention should shift to learning a variety of skills (Thompson & Rudolph, 1992). According to Erikson, a child who does not at least partially accomplish the tasks of one stage cannot move to the next. Thus, to understand the origins of a child's failure to undertake the skills of second grade, the counselor may find it useful to explore that child's earlier experiences.

Finally, it is important for the counselor to think about the fact that children have limited freedom to change the conditions of their lives and little experience in devising solutions to life's problems (Nelson, 1979). Typically, children cannot choose to live with different parents, to attend a different school, or to change the environmental influences that result from being born into a particular family and culture. "Children are reactors to changes in their living situations rather than initiators of change. They have relatively little power to take action to eliminate or prevent environmental causes of stress" (Prout, 1989, p. 18). Although freedom to make choices will gradually increase, children must often learn to live with things as they are. And because the child has had little experience with life, his or her store of strategies and skills is limited. Nelson (1979) said, "There are a great many ways in which young children are incapable and feel that incapability. There are things they cannot reach, words they cannot read, tasks they cannot perform" (p. 289).

In summary, it is important to realize that children are not simply small adults. Their behaviors are not mediated by thoughts in the same way as those of adults, and they have life tasks to perform that are shaped by their dependent status and their need to acquire skills. Among the skills that remain undeveloped is the use of language. Right and wrong do not carry the same meanings as for adults, or indeed across the different ages of childhood. Children are less capable of acting on their own behalf because of limitations of skill and freedom. And yet young children do experience emotional distress, feelings of inadequacy and endangerment, frustration, and many other problems that mirror their more mature counterparts. The remainder of this chapter addresses ways in which the counselor can help children even though they do not possess all the capabilities of older clients.

COMMUNICATING

As with any client, the first counseling objective with a child is to establish communication. It is important for the counselor to learn how the young client views and interacts with the significant others in his or her world. It is necessary to find out whether or not the client believes that he or she is loved, competent, and attractive. The counselor must adjust counseling procedures to maximize the client's ability to communicate experiences, taking into account cognitive, emotional, physical, and behavioral immaturity.

Confidentiality

In order to encourage a free flow of information, it is desirable to discuss confidentiality early in counseling, even with young children (Barker, 1990; Thompson & Rudolph, 1992). Many counselors become confused about the issue of confidentiality with children because of the dependency relationship that the client has with parents and parents' "right to know" what is happening with their children. In fact, the criteria for breaking confidentiality with children parallels that with older clients: confidentiality must be suspended if the client discusses actions that involve danger to self or others. Because of the child's inability to move about the world safely alone, running away is added to suicidal ideation as threatening behavior. Counselors are also required to seek intervention when a child reports instances of abuse.

Barker (1990) makes a useful distinction between two types of information that come from a counseling session. One type has to do with the child's condition (is he or she depressed, developmentally delayed, etc.). The other type of information is what the client actually told you or revealed in other ways during the counseling sessions. The first type of information is usually discussed with the parents and may be shared with teachers or others who are responsible for assuring a favorable climate for the child's growth and learning. Decisions about how much of this kind of information to share with whom should be based on considerations about the recipients' need to know and their ability to use the information for the client's benefit. The second type of information—the actual content of the counseling sessions—can and generally should be held in confidence. In instances when it seems desirable to share some of what actually transpired in a counseling session, the counselor should ask the client to help determine exactly what will be communicated and how. When either type of information is to be shared, giving the child as much involvement as possible in the decision about what to share and with whom maximizes his or her feeling of confidentiality and thus safety in the counseling setting.

Talking with Children

Young children are not oriented to approaching a counselor with a statement such as "Here is my problem" or "These are some of the things that have been happening to me." The counselor is initially an adult stranger, and many children are unaccustomed to talking with adults other than family members, close friends of the family, or teachers. Some children have never talked to an adult who really listens and tries to understand. Since most children have never experienced counseling, the counselor must structure what occurs in such a way that the young client understands what to do and is motivated to participate.

Many adults (including counselors who have not worked with children) are accustomed to talking *to* rather than *with* children, and even when they are trying to express concern and caring, they may be subtly condescending. We are reminded of a scene created by humorist Jean Kerr in which a doting relative walks into a room and exclaims, "My, look who is in the playpen!" the child is purported to be thinking, "Who do you *think* would be in the playpen?" All too often adults seem uncomfortable in a child's presence; they don't know what to say, so they comment on how the child has grown or changed. But growth and change are an expected occurrence in childhood, and children are often embarrassed by these awkward attempts at communication. Other people seem to believe that they need to talk loudly to children—who in fact have better hearing than adults as a rule.

It is best for the counselor to begin counseling with a child as with any other client—holding respect for the child's dignity as a human being and maintaining an attitude of exploration. The initial goal is to learn to know the individual. The counselor should speak in a well-modulated voice with a friendly manner, keeping in mind the vocabulary of the client and his or her level of cognitive maturity. The counselor should keep sentences short, use the client's terms, and use people's names rather than pronouns (Garbino & Stott, 1989). The task of the counselor is to enter the child's world and understand his or her experience from his or her perspective. This is the same as entering the world of any client during the initial-disclosure and in-depth exploration stages of counseling, with the important exception that the counselor must now reach back into the world of childhood to establish empathic contact.

Thompson and Rudolph (1992) recommend frequent use of "Tell me about . . ." as a way of encouraging expansive responses. "Tell me about your family" or "Tell me about school this year" will often result in stories that include both the happenings of importance to the child and his or her feelings about those happenings. Nelson (1979) stated that children benefit by hearing their feelings restated by the counselor, and that they need even more help than adults do in labeling their feelings. Thompson and Rudolph also caution: "Children easily fall into the role of answering adults' questions and then waiting for the next question; thus the pattern of a question and answer session is set" (p. 42). If an inquisition is allowed to begin, the counseling may self-destruct because the counselor loses the role of the listener and instead must devote all of his or her energy to generating more questions. "Counselors learn more by listening and summarizing than by questioning" (Thompson & Rudolph, 1992, p. 43).

Barker (1990) suggests two additional techniques for encouraging children to engage in revealing talk. The first taps into the client's fantasies by asking him or her to state three wishes. If one of the wishes is, for example, that Mommy and Daddy would not fight so much, the significance for the child's feelings of security is obvious. The second is to inquire whether the client ever has dreams. The majority of children acknowledge that they do and can then be encouraged to describe their dreams. If a client says that he or she does not dream, Barker sometimes uses the alternative of asking the client to make up a dream. Either of the dream techniques provides access to the child's private and even unconscious world.

Even though many children have not developed a lot of facility with language, it is usually possible to establish some conversation if the counselor remembers what childhood is like and expresses a genuine interest in learning something of the client's world. In fact, Barker (1990) states that some children—those who are older, of above average intelligence, and with good verbal skills—enjoy conversation and engage eagerly in telling their stories. However, many other children are more at home with play and may become acclimated to counseling more readily if toys and other media are provided. The next subsection provides a basic introduction to using play to encourage children's communication.

Use of Play in Counseling

The use of play and art supplies within a counseling session with a child will usually aid the communication process. Play has been linked to a number of cognitive phenomena, including problem solving, language learning, creativity, and the development of social roles (Brady & Friedrich, 1982). Therefore, observation of play offers a window on a child's thoughts and feelings. A little girl who is angry at her teacher can use crayons to draw her angry face or cast her teacher as dead in doll or puppet play. A little boy who is experiencing disorganization and abuse in his family can use a dollhouse and human figures to illustrate family life as he knows it. A child who is nervous and full of energy can let off steam by punching a punching bag. Given some things to play with, a child does not need to be told how to play. The counselor needs only to know how to encourage significant play, to observe what happens, and to talk with the child about it.

The introduction of play into counseling sessions with young children provides a means for communication that is natural for them. While playing, catharsis takes place: the child's nervous energy is expended in using a punching bag; the child's emotions can be expressed in a drawing; or the dynamics of the child's family become apparent through what happens in the dollhouse. The play provides a means for structuring time and a comfortable atmosphere in which neither the client nor the counselor feels the obligation to maintain constant conversation.

Nevertheless, conversation about the play helps the counselor understand the meanings the child attaches to what he or she does. In time, as the client gets used to being with the counselor, he or she is likely to engage in more and more direct conversation that extends beyond the play activity. The counselor's participation includes a physical presence, participation in the play, observation of activity, and verbal interaction with the client. As with older clients, early counselor responses are nonintrusive (minimum degree of leading), and the length of lead can be extended as the relationship builds.

Brady and Friedrich (1982) have suggested four levels of intervention: physical presence, reflecting or paraphrasing the child's statements, third-person interpretations, and direct interpretations. In the first level, physical presence, the counselor limits his or her responses to encouraging or describing the child's explorations and play: "Let's see what's in here" or "The boy doll is hiding under the bed." At the second level, reflection of feeling might extend to a statement like this: "The boy doll is really afraid of something." The third level, third-person interpretation, is a description of what the child seems to be experiencing without any attribution of the feelings involved to the child: "Sometimes it is very frightening if a child is left all alone at home for a long time." The fourth level is the most direct, attributing what has been displayed in the play to the child's actual experience: "You feel very frightened when you are left alone at home." It is evident that a fourth-level response probably has the greatest potential for moving to discussion of the child's real-life concerns. But it is safer for the child to begin talking about his or her concerns as though he or she were discussing someone else. A child who is acting out a scene where the mother is intoxicated may reveal much of what happens in his or her own family if the counselor simply comments on the experiences of the play figures. If the counselor prematurely says, "You get very concerned about your mother's drinking," the client may very well deny that his or her mother drinks and may become angry and reluctant to continue with the counseling.

In general, children probably do not reach sophisticated levels of insight about their personal problems through discussion of their play. As we indicated earlier, their cognitive skills, moral judgment, and experience all reflect immature equipment for coping with life. Nevertheless, catharsis—combined with exploration and problem solving through play—creates perceptual and behavior change, and the experience of being truly valued and listened to as a client helps build feelings of self-worth.

Some play materials are much more useful than others for promoting affective expression. Unstructured materials allow the young client to place his or her own meanings on the play and are therefore much better than structured games. Lebo (1979) systematically studied several different kinds of play material to determine which prompted the greatest amount and variety of verbal expression by his clients. His brief review of the literature indicated considerable support for the view that children who talk during their play sessions make more rapid progress than those who do not. He found that a dollhouse, family, and furniture were most effective in eliciting conversation with his sample of average-intelligence boys and girls, aged 4 through 12. Poster paints, brushes, paper, easel, and jars were nearly as good. Next in order, but somewhat less helpful, were a sandbox, chalkboard and colored chalk, cap gun and caps, coloring books, and hand puppets.

Dolls and puppets should be simple enough that the young client can use his or her imagination in playing. Among the dolls and puppets, there should be males and females and adults and children of various racial backgrounds. The child can use them to build the drama. Art supplies are valuable because they can be used to create anything the child needs to share.

We do not recommend using structured competitive games in counseling with young children. In the first place, such games require the counselor to enter into competition with the client, and there must be a winner and a loser. In most instances, if the counselor plays at his or her skill level, he or she will win, creating an experience of failure for the client. If the counselor lets the client win, there is the risk that the client will recognize the manipulation; then the genuineness of the relationship will be compromised. Second, because the rules of the game tell the players what to do, there is little opportunity for the client to infuse the interaction with material from his or her life. Although structured games have the initial advantage of instructions that tell one how to get started and although they provide a means for being with the child comfortably, these advantages are soon lost in the limitations that the structure places on communication. There are, however, some noncompetitive structured games designed for use in counseling, which function somewhat like incomplete sentence blanks by suggesting common concerns of children around which a story can be told or acted out. These may be successful with some children and are fairly easy for the counselor who is inexperienced in working with children to use.

It is appropriate for a counselor who is using media with a child to also engage in direct conversation about the daily events of the child's life. The client will become more comfortable with direct conversation as the two people spend more time together, and such conversation can be interspersed with conversation about the play that occurs.

We have found that materials can be useful in building communication with clients of all ages. As children mature and increase in their verbal and cognitive abilities, more and more of the counseling is likely to be direct conversation. Even with adults, however, it is sometimes helpful to provide a pen and paper and suggest that they draw how they feel about a situation they are having trouble verbalizing. As children approach puberty and begin to leave toys behind, puppets and dolls will gradually drop out of the sessions. However, pillows can still be punched to express anger, and drawing or sculpting may be a very useful activity. Verbal role playing will gradually replace doll and puppet activities as a way of bringing interpersonal situations into the counseling room.

Summary of Communication Techniques

This section has focused on how the counselor encourages communication with a child about things that are important to the child and that the child introduces into the sessions. Direct conversation accompanied by observation and interaction about play are the two primary methods. An attitude of respect for the child as an individual underlies any successful interaction, and language should be tailored to the child's developmental level. Leads should be fashioned to encourage expansiveness of client responses rather than narrow answers to counselor questions; play materials should be selected for their value in promoting imaginative play that reflects real-life concerns. The counselor should construct the experience so that he or she can spend time listening and observing. Counselors, of course, play a role in assessing and socializing children as well. Subsequent sections of this chapter address these topics.

ASSESSING

The assessment question when working with a child is basically the same as with any other client: "How is this person doing in meeting the challenges of his or her life?" But the assessment process with children requires special knowledge and skill because the capabilities of children change markedly from year to year and the challenges of life depend in large measure on the significant adults in the child's life. The basic assessment question is therefore adapted: "How is this person doing in meeting challenges for a person of his or her age, compared to what the adults in the environment expect?" Problems may occur because the child is in some way deficient in meeting age-appropriate challenges (such as assuming increasing responsibility for dressing himself or herself) or because adults are making unreasonable demands (such as demanding periods of concentration beyond the developmentally related attention span of the child).

The purpose of assessment, which should be done upon referral, is to begin setting goals and devising a treatment plan. As when working with adolescents and adults, the counselor attempts to identify arenas in which the child is not functioning effectively and to determine what factors might be contributing to the difficulty. With a child, it is often a significant other who is an adult who asserts that a problem exists, and the child may or may not perceive that there *is* a problem. In working with children, the counselor must take more responsibility for assessing the nature and source of dysfunction than with older clients. The counselor then works to set goals with the client or with significant others who affect the client's life.

Assessment Tools

A number of different procedures may be used in the assessment of a child's functioning:

- The counseling interview
- Observation of play in counseling sessions
- Observation in naturalistic settings
- Interviews with parents, teachers, and others
- School records, including anecdotal comments
- Incomplete sentence blanks
- Structured (clinical) interviews
- Tests
- Analysis of children's drawings

Use of all of these techniques can be enhanced through specialized training, but a counselor who has mastered the generic counseling process as presented in this book should be able to utilize the first six techniques at an initial level. We urge anyone who will work regularly with children to engage in training on all of these techniques that goes well beyond the coverage of this book.

As with older clients, a principal source of information used in the assessment process is what the child says in counseling sessions that are designed to encourage disclosure and exploration. However, since the young client's cognitive and language skills are less well-developed, the counselor should develop experience in encouraging play and understanding the language of play, as described in the preceding section. In addition to observing play in the counseling room, the counselor may also choose to observe the child in a natural setting (such as a child care facility or classroom). Such observation provides an opportunity to observe how well and how long the child is able to concentrate on assigned tasks, his or her success with assigned responsibilities, and his or her interactions with peers and authority figures.

Interviews with parents, teachers, siblings, and other adults who may have responsibility for supervising the child are a rich source of additional information. The inclusion of observations from several different people increases the total amount of the child's behavior that has been observed. Comparison of the views of different observers provides information about the perspectives of the observers as well. Direct discussion with parents and teachers about their expectations of the child helps complete the diagnostic picture.

School is one of the major responsibilities of school-aged children, and there is a great deal to be learned by reviewing the cumulative record. It is possible to determine whether the child is progressing satisfactorily, whether there are apparent strengths and weaknesses in different subjects, whether achievement seems congruent with measured ability, and how grades relate to results of achievement tests. Anecdotal entries in the record may provide clues as to how the child gets along with other children and staff.

Incomplete sentence blanks are structured written protocols, with stems of sentences to be completed by the client (for example, "I get frightened when . . ." or My mother . . ." or "A favorite time was . . ."). A child who is old enough to do so may write out responses, or the counselor may read the stems and record the client's responses. Inspection of the client's responses reveals areas where tension (or satisfaction) is expressed and provides an opening to discuss the issues identified. Many counselors find this technique useful in establishing conversation during the initial-disclosure stage of counseling, and little special expertise is required of the counselor who uses this means to stimulate discussion.

Clinical interviews with children and parents can be accomplished with the aid of published questionnaires such as *The Diagnostic Interview for Children and Adolescents* (Herjanic & Reich, 1982). These seek to identify particular symptoms and their frequency, duration, severity, and so forth. These protocols are usually used as a part of an intake process in clinical settings, and there is often a form for the child and another for the parents. The use of this technique has increased in recent years (Nystul, 1993), paralleling the increased use of the diagnostic categories for children and adolescents in the *Diagnostic and Statistical Manual of Mental Disorders* (American Psychiatric Association, 1987).

Use of standardized tests and analysis of children's drawings are special areas of study beyond the scope of this presentation. Usually some basic observations about drawings, including their completeness, facial expressions, and so on, can be made by a counselor who has a basic knowledge of child development. However, there is an extensive literature (Stabler, 1984) on children's drawings and their interpretation, and specialized training greatly increases the counselor's ability to use drawings for assessment purposes.

Scope of Assessment

The assessment process should focus on the child's functioning in each of the important sectors of his or her life, using the typical performance of persons of the same age as a standard of effectiveness. One among many examples of multimodal assessment models is BASIC ID by Lazarus (1981, 1990). The acronym BASIC ID stands for seven areas of functioning that may become problem areas treated in counseling: behavior, affect, sensation/school (e.g., headaches, school failure), imagery (e.g., self-concept, dreams, fantasies), cognition, interpersonal relationships, and drugs/diet.

Knowledge of age-appropriate functioning in each of these areas may be acquired through the study of child development and will become more familiar with experience. A convenient brief summary of many age-normed characteristics was developed by Muro and Dinkmeyer (1977). Their presentation includes two- to three-page descriptions of each age group from 5 to 12, complete with some implications for counseling arising from the developmental level of the child. The content covers a broad range—from a recognition that 5-year-olds are comforted and encouraged by structure but have only a twenty-minute attention span to the recognition that 12-year-olds are developing sexually and need the kind of understanding and information that can come from group counseling. However, no simple guide is a substitute for more extensive study and experience.

Children are born with different talents and capacities and to families that vary in their abilities to provide nurturance and to teach the ways of the world. This leads to differences in accomplishments and satisfactions as children grow. In addition, parents have very different expectations for their children that reflect the parents' own needs. The majority of children who come to the attention of counselors are children who are not progressing well, because of the chances of nature and nurture. Most of these children are experiencing normal, even expected, difficulties with developmental tasks, and most respond to the caring attention and environmental manipulations that counselors can accomplish. Just as with older clients, there are a few children who are experiencing such severe difficulties that specialized attention is needed to diagnose and treat the problem. Counselors should be alert to signs of child abuse, childhood depression, and severe forms of antisocial behavior that may lead a child to harm self or others. Consultation with and/or referral to specialists should be employed when these conditions prevail.

Case: Ryan

Ryan, age 7, was referred to a counselor at the university because his behavior was disruptive at school, he was not accomplishing his schoolwork, and he was disobedient at home. He was brought to counseling by his mother, age 27, an articulate and concerned single parent who worked long hours. Ryan's parents were separated, and Ryan and his mother constituted the household. Ryan's father had been at home until about a year prior to the referral, but Ryan saw him only infrequently after the separation. There was no history of family violence, nor did the parents openly carry on their personal conflict in Ryan's presence.

The initial interview revealed that the mother had very high expectations for her son. When school was over each day, Ryan was required to take two public buses to get from the parochial school he attended to his home. He let himself into the house with a key and was alone for three to five hours before his mother arrived home. During that time, he was expected to do his homework, clean the house (including dusting and vacuuming), and make initial preparations for dinner. After dinner, he was required to help with the dishes, and then he was often alone again while his mother entertained her boyfriend.

The mother felt that she was fulfilling her obligations to Ryan to the best of her ability. She worked long hours to provide for him and to pay tuition to a good school. She couldn't understand why he was so resistant to doing what she expected of him and felt panicked when Ryan started acting out in school and refusing to do his work. An intervention by the school psychologist, isolating Ryan when he misbehaved, seemed to result in more rebellion. The mother's concern for Ryan was demonstrated by her seeking help at the university, even though this added an additional activity each week to an already busy schedule.

In a client-parent-counselor session, the counselor observed that many of the mother's communications to Ryan were critical and disapproving and focused on very small issues. She chastised Ryan for fidgeting during the session, even though he was attentive and his behavior was reasonable for a 7-year-old boy. She repeatedly corrected his speech, using overly precise and formal language to do so. Discussion revealed that the mother and son had no "quality" time together: they did not talk with each other except about life maintenance activities, they did not read together, and they did not go anywhere for recreation. Furthermore, Ryan was not permitted to have friends in when his mother was not home. When she was home, it was too late. Thus, Ryan rarely played with children his age—for that matter, he rarely played at all.

When Ryan met with the counselor alone, he was initially reluctant to say much but was very compliant with the counselor's instructions. He appeared to like the full attention of an adult. As time passed, the counselor gave Ryan some crayons and paper and asked him to draw scenes of events in his life. Soon Ryan was picturing himself as a very lonely child with only demanding adults in his environment. He was very reluctant to express negative feelings about these adults, but finally began to tell how mean James, his mother's boyfriend, was. James did not physically abuse him, but he was very firm in requiring certain behavior and was not "nice" like Ryan's father. As time passed, all the adults in Ryan's life were eventually described as demanding, rejecting, and not very affectionate. Ryan was enabled to express his anger about the way he was treated.

As Ryan talked more to the counselor about his angry feelings, his disruptive behavior at school disappeared. The teacher was persuaded to use some positive reinforcers rather than isolation—and within two months, Ryan was again doing well in school.

The problem at home was harder to solve. Ryan's mother had a difficult time accepting the suggestion that she should spend more time with Ryan doing things for fun. She was a busy woman and believed she was already doing all she could for him. She did come to realize, through consultation, that she was perhaps too demanding in her expectations about housework for a boy so young, and she eased up a bit on those demands. Ryan began to learn how to negotiate with his mother so that he could do some of his work when he wanted and could have time to himself part of each evening. He was permitted to visit friends after his mother came home in the evenings and began to get some playtime. Counseling was terminated at that point because the mother had got what she wanted out of it. Her son was doing well in school and was more manageable at home. The counselor made every effort to continue the relationship with mother and son in order to accomplish a more positive relationship, but further counseling was refused.

Discussion

In this case, a variety of communication techniques were used to establish contact with Ryan's world. He was given the full attention and respect of the counselor. This was a marked contrast to the communication style of his mother, who was the most significant other in his life. Drawing was a comfortable activity that created material for projective understanding of Ryan's interactions with his mother, her boyfriend, his teacher, and other children. Physical activity was provided to permit release of energy. And, finally, comfortable verbal communication increased as counselor and client came to know one another and built a trusting relationship.

The assessment process revealed that physical demands were being made on Ryan that he could not meet. Socially, his interactions were all with controlling adults, the most important of whom (his mother) had little idea what to expect of a 7-year-old. He was not permitted to express his opinion on anything—compliance was the behavior of a "good" boy. Emotionally, Ryan was lonely and angry but, fortunately, not yet bitter. With some support, he was able to express his feelings of anger and seek ways of alleviating some of his loneliness. His learning had been disrupted by physical and emotional overload, but he was a bright child who easily readjusted to school when other factors were moderated.

Note that the primary goals of counseling were to get Ryan to express and unload some of his negative feelings and to seek some ways of adding companionship and variety to his life. At the same time, work with his mother was directed toward encouraging her to express love, to reduce the demands she placed on Ryan, and to try to understand what his life was like. These goals emerged directly from using the kind of assessment process described in this section.

Questions for Further Thought

1. What factors contributed to Ryan's anger?
2. What elements of the mother's personality formed the fabric of a negative living routine for Ryan? Which of her personality attributes were helpful to Ryan?

3. Do you believe that counseling resulted in satisfactory outcomes for the client? What else might have been attempted? How feasible do you believe these additional goals might have been?

SOCIALIZATION

Many of the difficulties that children experience are the result of deficits in experience. Remembering that a baby is born a largely unsocialized being a short five years before he or she first enters school brings into perspective how much growth and learning occur in a very short time. It also clarifies why it is so likely that some valuable learning may get overlooked in the life of a particular child.

A child such as Ryan in the preceding case study has had very little opportunity to learn the social skills involved in interacting with others of the same age. Other children, whose parents require very little of them, may develop deficits in such basic things as personal care and hygiene, to say nothing of more complex tasks. Children whose families are impoverished often have little contact with the world beyond a four- or five-block radius from their homes—territory that is usually impoverished, unvaried, ugly, and possibly dangerous. Children in these neighborhoods experience a limited range of role models, many of which are negative.

One advantage of working with young children in counseling is that such deficits are fairly easy to moderate if they are identified early. Sometimes small interventions such as helping a child learn how to come to school clean will pay big dividends in his or her relationships with others and in building self-esteem. A counselor can also serve as a resource to parents and teachers who are looking for ways to enrich the lives of children.

Case: Robina

Robina was an 11-year-old African-American girl who was attending a predominantly white, affluent, suburban junior high school. She was an average student but was regarded as a "curiosity" by students and faculty alike. She used foul, openly sexual language with everyone, and her appearance was often bizarre. Her use of make-up made her look like she had on a mask.

A female counselor at the junior high school decided to try to help Robina. Soon the counselor discovered a huge deficit in Robina's socialization experiences. Since birth, she had lived with her father in a small run-down dwelling at the edge of the community. He was in his late sixties and seemed to have no means of support. He seemed to care for Robina and provided for her to the best of his ability. However, he obviously was not equal to the task of guiding a pubescent girl. (It is hard to say why the school system had never provided any special services to Robina, except that she was an average student who caused no trouble until she began to mature physically.)

Robina very quickly began to relate positively to the counselor (who was an attractive and well-dressed woman), and her hunger for information was enormous. Many of her early questions focused on "what is beautiful." She seemed to have almost no aesthetic sense of what looks good in clothing or make-up. She also seemed confused because of her brown skin, which the other girls in school did not have. The counselor spent several sessions using black-oriented magazines to help Robina establish some sense of how she could look, and the girl's appearance changed practically overnight. She was highly motivated to look good, and with the help of some donated clothing and some information, she was able to look much more attractive.

Not surprisingly, her language also reflected a knowledge deficit. She had little sense of the meaning of some words she used and had no overt desire to be sexually involved with the people in her school environment. As she learned from the counselor the meanings others attached to what she said, she quickly began to refrain from lewd comments that had made her a laughingstock in the school.

Fortunately, Robina was helped at a crucial time and was able to continue with high school and develop a degree of acceptance among her peers, even in a school where she was racially different. She left school with the ingredients for a productive life.

Questions for Further Thought

1. The reasons for Robina's information deficits rested with her father's inability to provide needed information. Do you think this is a common problem? To what extent is it limited to parents who have suffered the effects of poverty?

2. What strengths do you observe in Robina?
3. Try to place yourself in the role of the counselor. What feelings might you have had in trying to relate to Robina's unsocialized and seemingly crude behavior? What knowledge might you have needed that you don't possess now?
4. In addition to the magazines that were used to talk about grooming and fashion, what other materials might you have used with Robina?

WORKING WITH PARENTS

Earlier sections of this chapter have established that children are profoundly affected by the adults in their lives, particularly their parents. The traditional family is changing, and about half of the children born in the past ten years will spend some time in a single-parent family (Hetherington, Stanley-Hagan, & Anderson, 1989). Even so, because of their dependent state, children are affected by the decisions of their parents, especially any decision that parents make about the make-up of the family unit. In many instances, it may be difficult or impossible to help a child change without enlisting the support and cooperation of the natural parents, custodial single parent, and/or stepparents. We will use the word "parent" in this discussion to mean those adults who are serving the parenting role for a given child, regardless of their exact relationship to one another or to the child.

The initial contact with parents may occur in a variety of ways. Sometimes a parent will recognize the need for help with his or her parenting behavior and will consult a counselor, ready to consider change. Sometimes a parent will contact a counselor and request that the counselor change the child in some way, but will not take any responsibility for the child's problem. At other times, particularly in school settings, a counselor will first have contact with the child and will decide to involve a parent only after having acquired the child's perspective on his or her interactions with parents.

Regardless of how a parent arrives at a counselor's office the first time, he or she needs to be respected and cared about just as surely as the child needs such understanding. With few exceptions, parents care about their children and are trying to be good parents.

The process of effective parenting is complex, and adult status alone does not assure competence. Few parents have any training in the art of parenting, and most have learned from their own parents some patterns of interacting that create problems. (It has been found that abusive parents almost always were abused as children themselves, though, of course, not *all* who have been abused in childhood end up as abusers.) Sometimes parents who might be able to provide a nurturing environment under favorable conditions are under stress because of work, money problems, health problems, or marital conflict. Under such conditions, the children's needs may be neglected, and frustration may be vented on them. In cases where stress exceeds a parent's capacity to cope effectively, emotional disturbance may significantly debilitate any parenting competence, and children may become fixtures in the parent's maladaptive lifestyle.

There are a number of procedures for helping parents improve their parenting effectiveness. Perhaps the simplest intervention is consultation with a parent on an occasional basis to encourage supportive efforts and learn about the progress a family is making toward positive interaction. Sometimes parent groups are established to provide training in parenting and to encourage parents to share their concerns and solutions with one another. Packaged training programs such as *Systematic Training for Effective Parenting* (Dinkmeyer & McKay, 1989) are excellent resources for counselors seeking to develop skills for consulting with parents. The focus of much parent *consultation* (education) is the improvement of parent-child and sibling relationships.

Counseling with parents moves beyond educational efforts aimed at improving the management of children and attempts to address some of the parents' own practical and emotional problems. Counseling interventions with parents can range from counseling about situational stress (e.g., caring for an ill and aging parent) to marital counseling to psychotherapy. Often the focus may move away from the children to debilitating problems in other sectors of the parents' lives. Although school counselors engage in extensive consultation with parents, they typically refer parents who need counseling to mental health facilities. Since the other chapters of this book provide an introduction to the counseling process, we focus in this section on contributions that consulting with parents can make to the counselor's work with children.

As indicated above, the first consideration in consulting effectively with parents is to accept each parent as a person in need of some kind of assistance, not as a "bad" parent. Second, it is important to remember that all children initially had two parents and most still do, even if divorce or separation has occurred. It is important to learn how each parent is functioning in the family. Third, parents who have suffered stress within the parenting process frequently become neglectful about expressing affection toward their children. Finally, parents often need help understanding what their children are trying to accomplish by their behavior so that they can help their children get what they want in more acceptable ways.

It is quite common for the initial contact to occur with only one parent, usually the mother, if self-initiated. If the child has significant contact with both parents, it is almost always desirable to request that both parents become involved in the consultation. Frequently, the two parents have differing perceptions of their child, and one may be defeating the parenting efforts of the other. The mother may see her son's behavior as unruly and aggressive, and the father may think he's "just being a boy." The mother may tolerate virtually any behavior, and the father may believe in strict control. One or the other parent may simply not pay much attention to what the child does. It is important to determine whether the parents are sufficiently reinforcing prosocial behavior and effectively punishing antisocial behavior (Patterson, 1982), while allowing sufficient latitude for normal experimentation.

Sometimes children become expert in playing one parent off against the other. It is best to see both parents to determine how much support each parent is currently providing to the child and what each parent's strengths may be. The counselor is being biased if he or she accepts the views of one parent and rejects or ignores those of the other. Families function as systems in which the behaviors of each member influence the behaviors of all others. Therefore, each parent needs the support of the other parent in order to initiate change in the family, and chances for success are enhanced if both become involved in the consultation.

Parents in blended families often have extra difficulty settling on their appropriate roles with the children. Whether the children were born to just one or to each of the adults, there is often the feeling that the birth parent has more rights and responsibilities with a given child than the stepparent does. Blended families work best when the parents become full partners in their interactions with each other and with the children. Otherwise, concerns about one parent feeling undermined by the other one and/or children playing one parent against the other can become extremely troublesome. It naturally takes time for parents and children to develop comfort with their roles when a blended family is formed, and many schools currently offer support groups for children with stepparents.

Many parents make a serious error in seeking to get their children to behave in desired ways by withholding expressions of their love for the child. The child, feeling hurt by the parents' apparent lack of caring, becomes resistant to the parents' wishes. This results in further withholding of statements of caring. In many families, no one can remember the last time someone said something caring to another. Ironically, such breaks in affectional communication are often based on caring: it matters to the parent what the child does; it matters to the child what the parent thinks. It is often very useful to point out to parents that someone has to break the standoff. Plans can be made to assure that both parents participate in reassuring the child of their love.

Finally, it is useful to help parents understand that children's behavior always has a purpose. Dinkmeyer and McKay (1989) suggest that children's misbehavior occurs to serve one of four purposes: seeking attention, seeking to control another (power), seeking to get even (revenge), or seeking to be excused from responsibility (withdrawal). A counselor can work with parents to help determine what the child wants and how he or she can be helped to make a place for himself or herself in a positive way. Through such intervention, often combined with continued counseling with the child, significant elements of the family system can be manipulated in support of the child.

In the consultation process, the counselor begins with the assumption that the parents want to improve the nurturing environment for their child and that they have the capacity to do so. (If the assumption of capacity proves false, then parent counseling or marriage and family counseling may be required.) The focus of the consultation process is usually on the parent-child and sibling relationships. Consultation helps parents to express their caring and to understand their children's motivations so that they can provide positive ways in which the children can fulfill their needs. Inconsistent and ineffective parenting behaviors are identified, and alternatives are planned, rehearsed, and reinforced.

SUMMARY

Counseling with children is based on the same three-stage process that is described in the earlier chapters of this book. However, children are developmentally different from adolescents and adults, and they are not as free as older clients to act on their own. The counselor must establish communication on the child's terms and enter the child's world of experience. This can be done effectively through the use of play media, and certain kinds of play materials tend to elicit verbal output from the child that helps in understanding his or her experience. Assessing the locus and extent of the child's problem requires an understanding of child development and can be enhanced by the use of such techniques as the clinical interview, incomplete sentence blanks, and tests. A child's dysfunction is often caused in part by lack of experience and deficits in abilities that are maturationally determined. Deficits can sometimes be reduced through socialization experiences that are planned as a part of counseling. In many instances, parents' help should be enlisted, and they often need consultation about how to be more supportive with their children while providing structure and setting expectations.

REFERENCES

American Psychiatric Association. (1987). *Diagnostic and statistical manual of mental disorders* (3rd rev. ed.). Washington, DC: Author.

Barker, P. (1990). *Clinical interviews with children and adolescents.* New York: W. W. Norton.
Brady, C. A., & Friedrich, W. N. (1982). Levels of intervention: A model for training in play therapy. *Journal of Clinical Child Psychology, 11,* 39–43.
Dinkmeyer, D., & McKay, G. (1989). *Systematic training for effective parenting: The parent's handbook* (3rd ed.). Circle Pines, MN: American Guidance Service.
Erikson, E. H. (1963). *Childhood and society* (2nd ed.). New York: W. W. Norton.
Garbino, J., & Stott, F. (1989). *What children can tell us.* San Francisco: Jossey-Bass.
Gilligan, C. (1982). *In a different voice: Psychological theory and women's development.* Cambridge, MA: Harvard University Press.
Herjanic, B., & Reich, W. (1982). Development of a structured psychiatric interview for children: Agreement between child and parent of individual symptoms. *Journal of Abnormal Child Psychology, 10,* 307–324.
Hetherington, E. M., Stanley-Hagan, M., & Anderson, E. R. (1989). Marital transitions: A child's perspective. *American Psychologist, 44,* 303–312.
Kohlberg, L. (1964). Development of moral character and moral ideology. In M. Hoff & W. Hoffman (Eds.), *Review of child development research* (Vol. 1). New York: Russell Sage Foundation.
Kohlberg, L. (1981). *The philosophy of moral development: Moral stages and the idea of justice* (Vol. 1). San Francisco: Harper & Row.
Lazarus, A. (1981), *The practice of multimodal therapy.* New York: McGraw-Hill.
Lazarus, A. (1990). Multimodal applications and research: A brief overview and update. *Elementary School Guidance and Counseling, 24,* 243–247.
Lebo, D. (1979). Toys for nondirective play therapy. In C. Schaefer (Ed.), *The therapeutic use of child's play.* Northvale, NJ: Jason Aronson.
Muro, J. J., & Dinkmeyer, D. C. (1977). *Counseling in the elementary and middle schools.* Dubuque, IA: Wm. C. Brown.
Nelson, R. (1966). Elementary school counseling with unstructured play media. *Personnel and Guidance Journal, 45,* 24–27.
Nelson, R. (1979). Effective helping with young children. In S. Eisenberg & L. Patterson (Eds.), *Helping clients with special concerns,* Boston: Houghton Mifflin.
Nystul, M. S. (1993). *The art and science of counseling and psychotherapy.* New York: Merrill.
Patterson, G. R. (1982). *Coercive family process.* Eugene, OR: Castalia.
Piaget, J., & Inhelder, B. (1969). *The psychology of the child.* New York: Basic Books.
Prout, H. T. (1989). Counseling and psychotherapy with children and adolescents: An overview. In D. T. Brown & H. T. Prout (Eds.), *Counseling and psychotherapy with children and adolescents* (2nd ed.). Brandon, VT: Child Psychology Publishing Company.
Stabler, B. (1984). *Children's drawings.* Chapel Hill, NC: Health Sciences Consortium.
Thompson, C. L., & Rudolph, L. B. (1992). *Counseling children* (3rd ed.). Pacific Grove, CA: Brooks/Cole.

Chapter 9

Issues in Counseling Women and Culturally Diverse Clients

The counseling profession has its roots in American and Western European culture of the nineteenth and early twentieth centuries. For many years, the clients who came to see professional counselors for assistance with problems were fairly homogeneous in culture and background. Moreover, they usually had the same cultural tradition as their counselors. In the late twentieth century, however, both the counselor population and the clientele became more culturally diverse, as a result of changes in the population and increased access to education and mental health care. Furthermore, demographic trends indicate that cultural diversity will be an enduring characteristic of the American people (Herr, 1989). In fact, women and minorities currently make up 75% of the entering work force, and by 2010, ethnic and racial minorities will constitute more than 50% of the U.S. population (Sue, Arrendondo, & McDavis, 1992). (Cultural diversity is also increasing in other Western nations, albeit at a slower pace.) Thus, it is clear that counselors will no longer serve only culturally homogeneous clients with whom they share a common background and that they must be prepared to expand their models of the counseling process to accommodate the needs of a multicultural society.

Furthermore, for a long time counselors and clients held similar and fairly clear definitions of the proper roles of men and women in society. In recent decades, as those rigid gender role definitions have been challenged, counselors have had to redefine their own values, examine their own biases about gender roles, and cope with clients' diverse perceptions about masculinity and femininity and experiences of gender bias and sexism.

This chapter discusses the impact of cultural diversity and gender on the counseling process. First, we examine why culture is such a critical influence on the counseling process. Second, we focus on the specific attitudes and knowledge base that counselors need for effective culture-sensitive counseling. Third, we identify several specific skills counselors should integrate into their counseling approach when working with culturally diverse clients. Finally, we discuss the special attitudes, knowledge, and skills necessary for successful counseling with female clients.

CONSIDERATIONS IN COUNSELING CULTURALLY DIVERSE CLIENTS

Why Is Culture Important in Counseling?

It is not uncommon to hear counselors say that they are "color-blind" or that "underneath, all people are the same." Such an attitude implies that culture is a relatively minor influence on human behavior and that the universality of human experience is the major influence. This emphasis on core human experience is termed an *etic* perspective. What usually follows from an etic perspective is a view that cultural heritage (either the counselor's or the client's) is much less important than the common human experience of the individuals who are interacting. The second consequence of an etic perspective is the belief that counseling theories and methods can be effective regardless of culture. Both new evidence from scholars and the experiences of culturally diverse groups in counseling reveal the flaws of this perspective. Culturally diverse clients, sometimes called minority clients, are less likely to seek out counseling services, are likely to be given more serious diagnoses than their majority counterparts, are more likely to drop out of counseling, and are less likely to have "positive outcomes" from the counseling they do engage in (Sue & Sue, 1990). In short, there is little evidence that the common human experience of people from different cultures is much more important than their cultural differences. Scholars of cross-cultural counseling argue that its relative ineffectiveness can be explained by the fact that an *emic* perspective on culture is more accurate (e.g., Pedersen, 1988). The emic perspective asserts that culture does indeed influence a whole host of aspects of human experience, including:

- The values one holds
- The language one uses and the nuances of meaning one attributes to words
- The use and interpretation of nonverbal behavior
- The definitions of what is normal and dysfunctional behavior
- The ways one seeks to get help for dysfunction
- The patterns of family and interpersonal relationships
- One's world view, or frame of reference for making sense of one's experience and set of deep assumptions about one's relationship with the outside world

One way in which culture impacts on the counseling process is by influencing verbal communication and the understanding of the language used. When counselors and clients are of different cultures, they may not be equally comfortable using English or they may speak different dialects of the language. Even if English is the first language of each participant, subtle differences in usage exist and there is no guarantee that nuances of meaning will be the same for counselor and client. Because verbal communication is the "currency" of counseling, language barriers can be an enormous roadblock to effectiveness. As discussed in Chapter 3, verbal communication is the primary means by which the counselor demonstrates empathic understanding. Without a commonality of language, empathic understanding is harder to attain and communicate to the client. How does a counselor get a true sense of a client's experience without a shared vocabulary? In addition, verbal communication is the dominant means by which the diagnosis of problems is made and the means by which intervention strategies are communicated and evaluated. When verbal communication is flawed, the risk of misdiagnosis and misapplication of counseling strategies is very high. Without shared language, then, counseling is hampered at every turn. After all, as Sue and Sue (1990) remark, "The presupposition is that participants in a counseling dialogue are capable of understanding each other" (p. 47).

Second, culture influences many nonverbal behaviors. Some nonverbals, such as smiles or frowns, seem to have universal meanings, but others are culturally relative. For example, in Euro-American culture establishing eye contact with another person during a conversation is viewed positively as a sign of self-respect, respect for the other, and attention to the conversation. Conversely, failure to establish eye contact during a conversation is seen as a sign of shyness, low self-esteem, or lack of interest in the partner or the conversation. In some cultures, however, whether one establishes eye contact depends entirely on the status of the other person and not on self-esteem or interest in the topic. In this case, when the other person has a higher status in society, to make eye contact is seen as aggressive or disrespectful to that person. For example, for a young person from some Native American cultures to make eye contact with an adult in authority would be seen as rude and presumptuous. In these cultures, eye contact is only appropriate between equals. Counselors from different cultural backgrounds who do not understand this influence of culture on nonverbal behavior are vulnerable to misinterpreting a Native American client's nonverbal behaviors. This simple misinterpretation of eye contact can lead to misdiagnosis of problems with self-esteem or assertiveness and misuse of counseling strategies such as assertiveness training.

Third, culture affects the behaviors a person perceives as normal or dysfunctional and the kinds of self-help strategies a person will use to solve his or her problems. For example, Sue and Sue (1990) suggest that for Asian-American clients, it is typical and normal for parents' desires for their offspring's career to take priority over the child's wishes. Counselors who do not understand the impact of this cultural tradition on Asian-American youth may misinterpret the young person's deference to the parents' wishes as lack of assertiveness or career immaturity. The likelihood of misunderstanding the client's reaction is greater when the counselor's culture emphasizes the "rugged individualism" that is a dominant philosophy in the Euro-American culture. As another example, in Asian-American cultures, a common self-help strategy for emotional distress is to distract oneself from the emotion or to ignore it; the assumption is that attending to it will make things worse. (Note that Sue and Sue suggest that this is a common approach in Asian-American cultures, not an approach that characterizes all Asian-Americans.) Again, counselors ignorant of this cultural message may misdiagnose a client's avoidance of emotional content as repression or lack of insight and may overuse affectively oriented interventions (Sue & Sue, 1990). The tendency to misunderstand the avoidance of negative emotions is stronger if the counselor's culture emphasizes affective awareness and exploration as the only path to mental health. Clearly, in some circumstances, avoidance or repression of negative emotions can be a serious impediment to healthy functioning. In these circumstances, attention to such emotions is a legitimate counseling goal even with Asian-American clients who share the view that Sue and Sue describe. The central concern of the counselor becomes sensitivity and skill in respecting the client's cultural tradition and possible difficulty with such an undertaking. The client needs to understand the rationale for such a focus and to freely choose it in spite of the attendant discomfort.

Fourth, culture affects family dynamics and interpersonal relationships. In Euro-American culture, the autonomy of the individual takes precedence over the family or the group, and many counseling interventions rest on this principle of individual freedom and responsibility. In many other cultures, however, the needs of the family or group take priority over the rights of the individual. Using the example of an Asian-American student's career dilemma, not only does the cultural tradition define deference to the parents' wishes as normal, it also suggests that open expression of disagreement with those wishes is improper and disrespectful. Moreover, discussion of family tensions outside the family is typically seen as dishonorable and disloyal. Obviously, this view of family dynamics will color the client's responses to the counselor's exploration of family conflict and the counselor's selection of strategies that include assertive behavior by the child toward the parents. Counselors who fail to take this cultural difference into account in assessing client needs or designing intervention strategies are likely to misunderstand family and group dynamics and risk an unsatisfactory or premature end to the counseling process. Clients with culturally insensitive counselors will see the counselor's suggestions as impossible demands that risk the dissolution of family bonds.

Finally, culture influences the world view of a person. Sue (1978) defined world view as one's "psychological orientation in life" (p. 458) and identified it as having two major components—locus of control and locus of responsibility. *Locus of control* refers to an individual's assumptions about his or her personal power to influence the events that happen. In Euro-American culture, a strong internal locus of control is assumed to be appropriate. An *internal* locus of control means that individuals see themselves as able to influence what happens to them, and an *external* locus of control places greater emphasis on the role of fate, luck, or historical accident in affecting human behavior. The latter orientation results in a more passive and accepting response to external events and is more common in other cultural traditions. Similarly, *locus of responsibility* refers to the placement of responsibility or blame for a person's plight, on the person or on the system or institution. Euro-American culture places the locus of responsibility squarely on the individual. Other cultures view the environment or social system as more responsible for individual outcomes. Furthermore, Euro-American culture has come to define mental health almost exclusively in terms of an internal locus of control and an individual locus of responsibility. The risk of a such a perspective is an underestimation of the real role that external events can play in a client's life circumstances and a minimization of the struggle involved in change and self-empowerment. A counselor with a narrow Euro-American world view would have a difficult time understanding or helping clients whose perspective on control and responsibility was at the other end of the continuum. Similarly, clients with an externally focused world view would be likely to see such a counselor's narrow focus on internal factors as blaming and blind to social and accidental influences. (For a more detailed discussion of the effects of culture on human behavior and the debate about an etic or an emic perspective, see the September 1991 issue of the *Journal of Counseling and Development* edited by Paul Pedersen.)

It is necessary that the counselor understand how the client's world view affects his or her situation and how it relates to the client's goals for counseling. Counselors ought not to judge a different world view as inherently inferior or problematic. While a counselor in the Midwest, one of us sometimes saw this mistake made with Native American clients whose world view tended to emphasize external locus of control and responsibility. Instead of understanding behavior in light of the world view of these clients, some counselors labeled their behavior as passive, resistant, or unambitious. Needless to say, counseling was not likely to succeed under these circumstances. Clearly, when clients present problems of adjustment to the mainstream society because of a different world view, the task of counseling may be to help them gain insight into the issue of differing world views. With this information, they can then decide how much they wish to change their behavior to function in the mainstream society. During this process, counselors need to help clients see how the different world views are operating and need to convey to them that the issue is the *difference in perspective,* without any suggestion that their world view is flawed or inadequate. Counselors with a Euro-American background should acknowledge that human experience can be viewed through several different lenses and no single lens provides an absolutely accurate view.

Self-Awareness: The First Step toward Cultural Sensitivity

To avoid what Wrenn (1962) called "cultural encapsulation," that is, the tendency to acknowledge only one cultural perspective or to define different cultures as deficient, counselors must first become fully aware of the influences of their own culture on their personal and professional behaviors. Understanding of his or her culture will give the counselor both a clearer sense of his or her cultural identity and the capacity to identify values, beliefs, and behaviors that are culture-specific and avoid the mistake of labeling them as universally true. This exploration of his or her cultural history can help the counselor move beyond cultural encapsulation, a critical need in this multicultural and often racist society.

Counselors should understand how they have acquired negative views of other cultures and identify what stereotypes underlie their attitudes. Confronting one's own racist and stereotyping attitudes is a painful endeavor, and the impulse may be to avoid it or to claim that one is "better than that" or "smarter than that." After all, counselors naturally want to see themselves as helping persons, truly devoted to the well-being of others. This impulse must be resisted because racist attitudes are insidious and often held unconsciously. Only commitment to their exploration will ferret them out and open them to change. Cultural elitism, the view that other cultures' values and practices are inferior to one's own perspective, must also be avoided because clients from oppressed groups are alert to subtle racist messages, a sensitivity developed as a coping mechanism in a prejudiced society. Only counselors who have truly understood their own cultural heritage and dealt with their own biases will be trusted by those who have had many other experiences of racist attitudes in those of the majority group.

One mark of a counselor who has made progress in this self-exploration is a higher level of emotional comfort with clients from other backgrounds. Another is a true appreciation for the richness and complexity of other cultural traditions and an openness to further growth in understanding one's own culture and those of clients. A detailed presentation of the counselor competencies necessary for culture-sensitive counseling is available in Sue, Arrendondo, and McDavis (1992). In addition, several of the books listed in the References at the end of this chapter will provide an excellent starting point for this process of self-exploration. It is important to note that this self-exploration is required of all counselors, not just majority counselors. Virtually all Americans have been acculturated in a racist and elitist society, and all counselors, whatever their race or ethnicity, will work with increasing numbers of clients with different cultural backgrounds.

Knowledge Base for Effective Counseling of Culturally Diverse Clients

The next step in becoming a culturally effective counselor is to become educated about other cultures. Where should counselors start this sizeable task? Counselors in Minneapolis will be likely to encounter clients of different cultural backgrounds than will counselors in Los Angeles or Toronto. Given current population patterns, for example, one would expect to find a greater density of Latino and Asian-American clients in Los Angeles than in Minneapolis, but a greater proportion of Native American clients in Minneapolis. Thus, the other cultures that a counselor needs to understand best will depend somewhat on the geography of the work setting. Population projections, however, suggest that all counselors will need to become knowledgeable about Latino and African-American cultural traditions because these cultures represent the largest, fastest-growing minority segments of the U.S. population.

To gain knowledge of other cultures, counselors should read the abundant literature on culturally diverse clients. (Again, the books on the reference list provide a good departure point.) Counselors need to focus this reading on understanding how each culture views language, nonverbal behavior, mental health and personality, and family and interpersonal dynamics. Counselors should also study the world view and values of each cultural group that is well represented in their service area. The multicultural counseling competencies proposed by Sue, Arrendondo, and McDavis (1992) also suggest that counselors become involved with the cultural group in the community so that their knowledge is more integrated and less academic. Counselors may attend festivals, political meetings, community events, and similar activities to develop this fuller sense of the culture. Counselors should be open to friendships with persons of different cultural backgrounds and actively seek out such personal connections. (The absence of friends from different cultural backgrounds may be one signal that a counselor has not yet succeeded in overcoming cultural elitism.) Such immersion also gives the counselor an appreciation of the contribution of the culture to the society, an appreciation that cannot be obtained in the classroom or counseling office. [See Lee and Richardson (1991) for more information on this topic.]

The second aspect of a knowledge of the cultural heritage of one's clients involves understanding how racism and cultural elitism have negatively impacted them. Sue, Arrendondo, and McDavis (1992) call for understanding of "immigration issues, poverty, racism, stereotyping and powerlessness" because these issues impact clients' responses to counseling, to the majority culture, and to their own cultural heritage (p. 482). Such awareness is particularly critical for counselors who have been fortunate enough to be spared such injustices; their background does not predispose them to understand the interrelationship between personal difficulties and the sociopolitical environment.

Third, there is a growing literature on cultural identity development that can be a valuable resource for counselors in understanding their own cultural identity and that of their clients. This literature posits that cultural identity development proceeds in levels or stages, just as others have theorized that personality development moves in stages. Atkinson, Morten, and Sue (1989) present a five-stage model of racial/cultural identity development, summarized in Table 9.1. This model focuses on identity development for persons whose cultural or racial backgrounds have made them members of a minority or oppressed group in a dominant culture. Atkinson, Morten, and Sue propose that cultural identity development is manifest in one's attitudes toward self, toward one's own cultural group, toward other minorities, and toward the dominant group. The lower stages are characterized by negative or conflicted views about oneself and one's culture and the same views of other minorities but more positive evaluations of the dominant culture. At these stages, there is some self-hate or self-depreciation, a negative view of one's own culture, and an identification with the values of the dominant culture. At the middle stage, one seems to move to the other end of the continuum, developing self-pride and an appreciation for one's culture while depreciating the dominant culture and holding conflicted views of other minorities. At this stage, the person comes to identify with his or her own group and thus has empathy for the minority role of other groups, but still views his or her own culture as superior. Very few of the positive attributes of the dominant culture or of individual members of it are acknowledged. At the higher stages, the person maintains the positive view of self and culture and begins to develop a less culturally elite view of other minority cultures. At these levels, a more differentiated view of the dominant culture emerges—with less depreciation of the culture as a whole and a more positive attitude, especially toward individuals who do not manifest cultural elitism.

Even though research evidence in support of this theory is meager and its hypotheses must be considered tentative until such evidence is accumulated, it is a useful heuristic for approaching counseling with culturally diverse clients. Atkinson, Morten, and Sue (1993) recommend that counselors assess the stage of cultural identity that best characterizes their client's views and respond accordingly. For example, for individuals at the lower stages, there may be little conscious discomfort with a majority counselor and little conscious rejection of the views of even a culturally encapsulated counselor. However, the negative views toward self and culture of origin need to be taken into account if empathic understanding is to be achieved. When a client is at stage three of cultural identity development, he or she may feel mistrust of a majority counselor. This client may respond with challenging or

Table 9.1 Racial/Cultural Identity Development

Stages of Minority Development Model	*Attitude toward Self*	*Attitude toward Others of the Same Minority*	*Attitude toward Others of Different Minority*	*Attitude toward Dominant Group*
Stage 1—Conformity	Self-depreciating	Group-depreciating	Discriminatory	Group-appreciating
Stage 2—Dissonance	Conflict between self-depreciating and appreciating	Conflict between group-depreciating and group-appreciating	Conflict between dominant-held views of minority hierarchy and feelings of shared experience	Conflict between group-appreciating and group-depreciating
Stage 3—Resistance and immersion	Self-appreciating	Group-appreciating	Conflict between feelings of empathy for other minority experiences and feelings of culturo-centrism	Group-depreciating
Stage 4—Introspection	Concern with basis of self-appreciation	Concern with nature of unequivocal appreciation	Concern with ethnocentric basis for judging others	Concern with the basis of group-depreciation
Stage 5—Integrative awareness	Self-appreciating	Group-appreciating	Group-appreciating	Selective appreciation

From Donald R. Atkinson, George Morten, and Derald Wing Sue, *Counseling American Minorities: A Cross-Cultural Perspective,* 4th ed.

nonrevealing comments to counselor inquiries and could be mislabeled as hostile or resistant if the counselor is unaware of the role of cultural identity in the interaction. Unless the counselor addresses that cultural mistrust, an effective counseling relationship may never develop. Sue and Sue (1990) suggest that it is possible that clients at this stage of cultural identity development may need more self-disclosure from the counselor about the cultural difference to allay their fears and mistrust. They suggest direct responses to client statements questioning the counselor's ability to be helpful, such as "It sounds like you're wondering whether I can really help you because I am white and you are Latino." Counselors in the midst of becoming culturally sensitive may also find this model a useful gauge to measure their own progress in this task.

Skills for Effective Counseling of Culturally Diverse Clients

Not only do counselors need more self-awareness, understanding of other cultures, and knowledge of models of cultural identity development, they also need a specific set of skills to work effectively with multicultural clients. Although the following is not an exhaustive list of the skills needed by counselors in a multicultural society, taken together, these recommendations describe fundamental competencies.

First, because a growing number of Americans do not have English as their first language, counselors who are bilingual will be able to serve their clients better. At the very least, counselors need to be aware of the burden of using English for some clients and be open to the inclusion of a translator (perhaps a family member) in counseling sessions. When language barriers are present, counselors should use methods that are somewhat less language-dependent (behavioral techniques or expressive media such as art and music).

Second, counselors need more flexibility in the means by which they convey the core conditions of unconditional positive regard, empathy, and genuineness. Ivey and Authier (1978) suggest that the core conditions are indeed universally appropriate, but that the methods typically used to convey those core conditions (reflecting, paraphrasing and letting the client lead the self-exploration) may be culture-specific. For a client accustomed to authorities who direct and structure interaction, for example, a narrow response repertoire that is limited to clarifying, reflecting, and summarizing client comments may be so uncomfortable to the client that he or she will not be able to tolerate the ambiguity. In such circumstances, the client may misinterpret the counselors' efforts to build a trusting relationship as lack of skill or knowledge. The flexibility to take into account the client's deep assumptions about the role of an expert counselor and the usual pattern of interpersonal relationships in the client's culture is critical to the success of the counseling enterprise. Specifically, in such a situation, counselors may need to engage in more extensive client orientation to counseling and greater use of structuring and questioning techniques. Some use of affective exploration responses is natural and appropriate, but exclusive reliance on them in the early stages of counseling may not be prudent.

Third, counselors need to be adaptable in their attention to intrapsychic issues and affective material. Sue and Sue (1990) point out that for some culturally diverse clients, attention to career and educational needs has a higher priority than self-exploration or affective functioning. Unfortunately, the folklore of the counseling profession—and the hidden agenda in many training programs—is that career counseling and educational counseling are more superficial or less personal forms of counseling. The message has been that "real" counselors attend to affective content and intrapsychic dynamics. Such a message fails to acknowledge the centrality of career and educational choices to personal identity and the enormous distress career and educational problems can produce. Counselors need to be sensitive to their own bias about these forms of counseling and recognize different expectations for the counseling process for some clients. Once this new attitude is achieved, counselors need skills in career and educational counseling to be able to work effectively in these domains. Similarly, there is some evidence to suggest that culturally diverse clients respond better to more structured, directive, and behavioral counseling approaches than to affectively oriented sessions (Sue & Sue, 1990). Although no prediction can be made about what any individual client needs or expects from counseling, counselors must be open to the use of more active and directive approaches and must have the skills to employ them.

Fourth, counselors should include intervention strategies that not only involve personal actions and feelings of the client, but also acknowledge institutional influences on the client's behavior or situation. For culturally diverse clients, many of their difficulties can stem from racist or prejudiced actions by specific others or by institutions. For example, an African-American woman may feel depressed and hopeless not because of low self-esteem, negative expectations for the future, or inadequate self-help methods, but because her boss may have actively impeded her promotion because of her race and sexually harassed her because of her gender. In this situation, a counselor must be open to identifying the central issues as external to the client's psychic functioning and assisting the client in responding to the injustices in her life. Some intrapsychic factors may be operating in her depressive feelings, but the counselor must see the complexity of causes in this case. Otherwise, the counselor may inadvertently seem to "blame the victim" (Ryan, 1971) for the racism and sexism she is experiencing. Instead, the counselor needs to help the client develop the skills and sense of personal power she needs to address these injustices.

Fifth, because counselors' knowledge of other cultures cannot always be complete and current, counselors need to develop contacts with professionals of diverse cultural backgrounds who are willing to consult with them to make the counseling process more effective. These consultants can also serve as referral sources in situations where the counselor's gap in understanding is too great or the client's preference is for a counselor with a similar culture. If such a consultant is not available to a counselor, that counselor has an added responsibility to seek out written material on the client's culture.

Because there is no such thing as a "culture-free" test, counselors also must be skilled in using standardized assessment instruments appropriately with diverse clients. Standardized tests that do not have norms for minority groups should be used with great caution. Whenever a test is employed, the counselor bears the responsibility for ensuring that the test results are accurate and are employed in ways to promote the best interests of the client. Counselors should also familiarize themselves with newer assessment methods that are somewhat less culture-dependent than earlier instruments.

Counselors need to match the intervention strategies they use to the needs of the individual and the cultural tradition that impacts most strongly on that individual. As we mentioned previously, methods arising out of the Western tradition that emphasize individual autonomy as the prime value and personal responsibility for one's actions as a core assumption must be used cautiously with persons whose cultures give stronger weight to family or group needs. In such situations, greater involvement of the client's family in the selection of counseling goals and strategies is critical for the success of counseling.

Finally, counselors need skills in working with the traditional support persons that are part of the culture and an openness to cooperation with designated helpers in the community. These traditional support persons include religious leaders, family matriarchs or patriarchs, or others whom the client sees as culturally accepted sources of help.

Cautions about the Emic Cultural Perspective

There is one danger of taking culture into account in the counseling process—the danger of stereotyping individuals. Not all Asian-Americans view family identically. Not all Native Americans resist eye contact with higher-status persons. The examples are almost endless. The point is that the individuality of the person is not lost in the culture. Cultural traditions must be seen as a hypothetical influence on a particular individual until evidence accumulates that they have been embraced by that person. For example, discussing the African-American culture in the singular form is really inaccurate. There are many subcultures within the broader African-American culture, and the diversity of experiences may be lost on a counselor who is overemphasizing culture or viewing it too narrowly. Thus, there is a balance to be achieved: culture is not to be ignored, nor is it to substitute for understanding the particular experience of the individual who seeks counseling.

Case: Lee

Lee is a 30-year-old architecture student who emigrated to the United States from Southeast Asia two years ago, leaving his family behind. He had been a licensed architect in Cambodia but his credentials are not recognized outside that country. He requests an emergency appointment at the university counseling service in the middle of the semester. As he enters the session, he appears outwardly calm and dignified, but his voice shakes and the expression on his face is pained. Three weeks ago, Lee received word that his mother became ill the day before she was to emigrate to the United States and then died one day later in the hospital there. His brother, the only remaining family member, is so lost in grief that he is saying that he no longer wants to come to the United States. Lee made the counseling appointment because he is having great difficulty studying and does not think he will be able to complete the mid-term projects that are due soon. He is worried that he will flunk out of school. He says that his purpose for emigrating is now lost, but since he can't return to Cambodia, he must find a way to carry on. His eyes become moist as he speaks, but no tears fall from them. He responds fully and respectfully to probes about his loss and recent experience but seems more focused on his academic concerns. After they have talked for several minutes, the counselor, a white woman, asks Lee whether he would be more comfortable discussing this painful issue with a counselor at the international student advising office. Lee immediately declines, indicating that he is an American now and no longer an "international" student.

Think for a moment about how you would proceed in this situation. How much do you know about the cultural background of this client? What role does the client's ethnicity and your knowledge of it (or lack of knowledge) play in your decision making? How do you balance his focus on academic issues with the reality of his grief and loss? What role, if any, do you think the gender difference between counselor and client plays in your decision making?

The approach we recommend in this situation depends, of course, on the counselor's degree of knowledge of the Cambodian culture. Since most counselors are unlikely to have much understanding of this tradition, we will focus on that aspect. When confronted by a client whose background is culturally different and little known to the counselor, the best course of action is a conservative one that respects the client's wishes and expectations for counseling. In this case, Lee was most pressured about his academic performance, and so counseling should first attend to this issue. The value of this emphasis is also indicated by his lack of experience with American universities and possible ignorance about the option of incompletes or withdrawals from courses because of medical and family emergencies. The counselor should first explain to Lee his alternatives for dealing with his courses for the semester and then discuss the advantages and disadvantages of each possibility. Practical information about how to handle the university bureaucracy to get course extensions may also be appropriate given his newness to the system and his level of grief.

What about all that affective pain? Is it enough to attend only to educational issues? This is a difficult question. On the one hand, Lee is already isolated, and an empathic listener may be especially valuable at this juncture. The possibility that his grief may deteriorate into clinical depression at some point is real and should be considered. On the other hand, he is an Asian man who appears to have a strong investment in acting with dignity and self-control and encouraging affective expressions of grief may leave him feeling ashamed and unwilling to show his face to this counselor again. Given the tendency of Asian cultures to have clearly defined gender roles, the gender of the counselor may also make this client feel uncomfortable about open expressions of grief. Thus, the latter argument is more compelling, we think, given the possible negative implications of a "loss of dignity" for Lee. The prudent course of action is to assist Lee immediately with his academic concerns and follow up with a second appointment to be sure those issues are resolved. The counselor should walk a fine line here, not cutting off Lee from expressing his pain but also avoiding a strong focus on the affective material. In the meantime, the counselor can consult with her resource person on Southeast Asian cultures and get some supervision on how to proceed responsibly with Lee's grief and pain. With this course of action, the client has also received what he asked for, is likely to view the counselor as expert and credible, and thus is more likely to return should the depression or academic problems worsen.

SPECIAL CONSIDERATIONS IN COUNSELING WOMEN AND GIRLS

Why should counselors consider women a population with special counseling needs? Some experts have argued that women and girls have unique ways of experiencing the world (Estes, 1992), ways of thinking (Belenky, Clinchy, Goldberger, & Tarule, 1986), making moral judgments

(Gilligan, 1982), and using language and processing information (Tannen, 1990). These scholars contend that this unique perspective is grounded in women's developmental experience and, perhaps, their biology. However, all of these theories are newly developed and, although they present intriguing hypotheses, need additional research before their validity is fully demonstrated. What is indisputable at this point, however, is the evidence that women and girls have experiences in American culture that give them mixed or negative signals about their gender. This narrow set of beliefs about what women and girls ought to be like regardless of their individual characteristics is called *sexism.* Specifically, women in this culture have been valued for the bodies, their reproductive capacities, and their domestic capabilities. The message in the culture is that women who are sexually attractive, able to reproduce and nurture children, and carry out domestic and caretaking functions skillfully are valuable. Conversely, women who do not hold these interests or attributes are seen as less worthy and usually dysfunctional. Obviously, not all men and women in this culture hold these views of women. The cultural message is common enough, though, that all counselors ought to examine whether they too have absorbed this narrow view of women. Similarly, counselors should be aware of the possible influence of these cultural messages on their male and female clients.

Research demonstrates that sexist attitudes in the culture have resulted in discriminatory practices toward girls in schools and colleges, which diminish their educational opportunities. For example, research suggests that girls are not encouraged to be skilled in math and science in school (American Association of University Women, 1992). Women in the workplace experience sexism through outright employment discrimination and the "glass ceiling," which interferes with their promotion to executive positions. In addition, a large body of evidence has accumulated that indicates that women in American society are at significant risk for victimization, an experience of physical, emotional, or sexual abuse by a person they know (American Psychological Association, 1990). The list of victimization experiences to which females are more vulnerable is long and includes sexual harassment at school or workplace; acquaintance, or "date," rape; physical and verbal battering by a spouse or significant other; rape by a marital partner; sexual abuse during development; and sexual exploitation by a therapist or other helping professional. For example, current statistics indicate that almost one third of adult women have experienced sexual abuse by a person close to them during their lives. (One of every six adult men has had an experience of sexual victimization during development.) Also, more than 90% of those who are sexually exploited in counseling and therapy are female clients of male therapists.

Other evidence reveals that the health problems of women have been ignored, underestimated, or mistreated by health care practitioners and researchers. For example, until 1992, funding for breast cancer research was lower than funding for several male cancers even though breast cancer has been a more common and life-threatening form of cancer for several decades. Research on the effectiveness of new treatments for common health problems such as heart disease usually excludes females in its trials. Thus, these treatments end up being used on women without benefit of any prior knowledge of how they may affect female physiology. In short, girls and women have been viewed and treated unjustly in Western society in a variety of ways, and counselors must expect that history of unfair treatment to affect the counseling process.

The following sections discuss how this cultural attitude toward women affects counselors' beliefs and values, the ways in which this experience impacts on women's response to counseling, and the specific skills the counselor must have to be effective with women clients. Finally, we will attend to the issues for women who come from culturally diverse backgrounds (and thereby deal with both sexism and racism) and to the implications of gender bias for male clients.

Counselor Attitudes toward Women Clients

There is a large body of research that has examined the degree to which counselors and therapists share the sexist views of the culture. The landmark study that initiated this research was conducted by Broverman and her colleagues (Broverman, Broverman, Clarkson, Rosenkrantz, & Vogel, 1970). They found that therapists had different definitions of what constituted a healthy adult, a healthy man, and a healthy woman. They reported that the therapists in their study used adjectives such as submissive, emotional, excitable, easily hurt, less independent, less objective, and less competitive to describe the healthy woman but did not use any of these adjectives when defining a healthy adult or a healthy man. Broverman and her colleagues characterized their results as indicative of a double standard of mental health for women:

> Acceptance of an adjustment notion of health, then, places women in the conflictual position of having to decide whether to exhibit those positive characteristics considered desirable for men and adults, and thus have their "femininity" questioned, that is, to be deviant in terms of being a woman; or to behave in the prescribed feminine manner, accept second-class adult status, and possibly live a lie to boot. (p. 6)

Although not all of the research to follow Broverman's study came to the same devastating conclusion about sexist attitudes, enough of a trend exists in the findings to cause experts to conclude that sexism of helping professionals, both male and female, is a continuing problem. Certainly, the rigidity of definitions of how women ought to behave has loosened since 1970, but the basic prejudice has not been erased entirely. The existence of such attitudes is not surprising since counselors have been raised in a culture that stereotypes women. Thus, the first step in effective counseling with female clients is the process of self-examination and self-knowledge that is also required for successful cross-cultural counseling. It is important to examine one's own development and explore the messages passed down about appropriate role behaviors for males and females. In some families, the sexist attitudes are overt and easily identifiable; in others, the messages are subtler and harder to detect. The following questions may be useful in the self-exploration:

Developmental Experiences

- How did your family react to activities of any member that were not consistent with the predominant gender role stereotypes?
- Did your family convey that appearance, nurturing capacities, and domestic skills were more important for women than men? Were girls in the family more often praised for looking pretty and boys more often praised for their accomplishments?
- How did the family react when dealing with a man or woman in a nontraditional role? Did any member of the family show discomfort or convey the message that there must be something wrong with the person for his or her career choice?
- What family messages were conveyed about emotional expression for males and females? For example, were boys discouraged from expressions of sadness, fear, or nurturing and girls discouraged from expressions of anger?
- Did males in the family get favored treatment in terms of financial support for education, especially higher education?

Current Experiences

- How much of that developmental experience is still influencing your behavior today? Are you still discouraging males from crying or females from open expressions of anger?
- How much do you expect females to take responsibility for nurturing others and exclude men from such responsibility? For example, when fathers care for their young children, do you tend to label that as babysitting instead of parenting, as you would see it for a mother's care of young children?

- How comfortable are you with persons in nontraditional career choices? If you get services from a plumber, surgeon, or nurse, do you respond differently if the person is male or female?
- If a woman remarks to you that she doesn't want to be married, is not interested in having children, and is not particularly concerned about appearing attractive, how does that statement strike you? Do you tend to think something must be wrong with her psychological or interpersonal functioning? Conversely, if a male indicates that he is not particularly ambitious, wants to place more emphasis on family than on career, and is sensitive and emotionally expressive, does that intuitively strike you as problematic? (Clearly, individuals who are denying any affiliative needs with other persons are not functioning at their highest possible level. A counselor should attend to such difficulties with intimacy regardless of gender. The point here is whether you tend to look at the tendency to evaluate affiliative needs differently for males and females.)

Once an honest self-assessment has been completed and a counselor knows the degree to which he or she has internalized the cultural stereotypes about gender, the counselor can begin the attitude change process. For some attitudes, change will come almost simultaneously with the recognition of the existence of sexism. The "aha" experience of uncovering the stereotype will go a long way toward bringing change. For other attitudes, a longer process that includes reading about sexism, discussing attitudes with colleagues and supervisors, and even seeking out consultation with a therapist may be useful.

The Implications of Sexism for the Client's Experience in Counseling

Unfortunately, sexist attitudes toward women are not limited to Euro-American culture. Such beliefs characterize many other cultural groups. Thus, although the specific form and expression of sex stereotyping are somewhat culture-specific, the existence of some rigid gender role expectations is fairly widespread across cultural groups. Thus, the great majority of clients who enter counseling have been raised in cultures with some gender bias, and they will be continuing to experience it in their interactions with the majority culture. There are two particularly important considerations about working with women in counseling. The first is that they are likely to have internalized to some degree the view that women who are attractive, nurturing, emotionally expressive, and domestically competent are more valued or valuable in our society. Second, they are likely to have experienced some form of discrimination or victimization because of their gender. (Note that these are probability statements; the experience of any individual female client may be different on either or both dimensions.)

The cultural definition of female attractiveness is frequently embedded in the concerns a woman brings to counseling, especially since the current standard for body shape can be attained by a very small minority of women. The thin body type with small waist and large breasts that characterizes models, actresses, and beauty contest participants is simply not genetically possible for most women. Thus, the standard against which women measure their attractiveness is as unattainable for most as running a three-minute mile. Yet the culture places so much attention on physical attractiveness that many women feel they must keep trying to attain the standard anyway and, in the meantime, devalue themselves for their lack of success. Recent evidence suggests that the focus on thinness and appearance is affecting girls at younger and younger ages. In the face of these cultural messages, it is not surprising that body dissatisfaction and low self-esteem are common problems that females bring to counseling. Even for those few women whose body shape matches the cultural standard, appearance may still be a subtle or overt counseling issue since they may wonder whether their appearance is the only aspect that makes them valuable to other people.

Many experts have asserted that this emphasis on appearance plays a causal role in the high incidence of eating disorders among women. Women who get caught in meeting the cultural standard for thinness are at greater risk for untested diet schemes, injury to their health from repeatedly losing and gaining back weight, and the more serious problems of anorexia nervosa and bulimia nervosa. In addition, body satisfaction is one part of a person's self-esteem, and the loss of such satisfaction because of unattainable cultural norms for women means that global self-esteem may be at risk as well.

Thus, women clients can bring to counseling a good deal of body dissatisfaction, a greater risk of eating disorders, and a sense of self-worth inappropriately tied to appearance and sexual attractiveness. Women may also use the same standards the culture uses in measuring the worth of other women and have difficulties in their interpersonal relationships with both men and women. Adolescent girls who are adjusting to their rapidly changing bodies may be particularly affected by these cultural standards.

The second major impact of sexism on women's experience is their high risk of an experience of discrimination or victimization. This fact is disturbing in itself, but experts also suggest that an experience of victimization early in life, such as sexual or physical abuse, leaves that individual at greater risk for future exploitive experiences. Several explanations have been presented for this phenomenon, but the essence of the arguments is that girls come to experience victimization as normal, unaware of how pathological such behavior is, and that women internalize the message they receive from their abusers—that the responsibility for the abusive behavior lies on the woman's shoulders. Thus, these women often carry enormous guilt and have low self-esteem and are therefore vulnerable to those who want to manipulate them. Frequently, memories of victimization are repressed, but the scars of low self-esteem and guilt remain.

Even if women have escaped victimization during their development, they may have experienced discrimination or sexual harassment at their workplace or school. For example, some evidence suggests that as many as 70% of women in the workplace have been the victims of sexual harassment (American Psychological Association, 1990). Although women who experience sexual harassment are less likely to blame themselves for their coworker's or boss's behavior, they still feel shame and embarrassment about the harassment and usually feel little power to make it stop. This sense of powerlessness can lead to depressed and angry feelings and sometimes to behaviors that seem counterproductive. Some scholars argue that victimization and discrimination experiences in which women feel powerless are major contributors to the high rate of depressive disorders among women (APA, 1990). Adolescent girls and women in Western societies have twice the rate of depressive disorders as adolescent boys and men. The risk of victimization can also lead many women to seek counseling for post-traumatic stress, either as an immediate response or as a delayed reaction.

In short, the victimization experience of females suggests vulnerability to very low self-esteem, high amounts of guilt and shame, and symptoms of depression or post-traumatic stress. These feelings often bring a girl or woman to counseling.

The cultural expectations about women identified first in the Broverman study cited above suggest that there are also more subtle impacts of sexism on women's experience of the counseling process. Since traits of submissiveness, influenceability, sensitivity to others' feelings, and lower levels of independence have been valued in women, female clients who have internalized these cultural values may have greater difficulty in disagreeing with a counselor, in striking out independently, or in taking action that they think may offend the counselor. In other words, traditionally socialized women may be at risk for too much deference to the counselor or too much attention to a positive relationship with the counselor, with the result that their counseling needs are not addressed. By the same token, counselors who have internalized these standards may be uncomfortable with assertive or independent women who do not work to nurture the relationship and may also be diverted from the important individual agenda by such stereotypes. In addition, counselors who expect women to be more emotional, more easily hurt, and less objective than men may also reinforce cultural stereotypes about women and may overlook opportunities to have a client process cognitive material or deal with material that may be painful but essential for future growth.

It is important to remember, however, that these are general issues likely to be present for many women who enter counseling but not necessarily present for any particular woman who enters a counselor's office. The unique experience of the individual may or may not fit with the experiences of women in general. Counselors must hold these ideas as tentative hypotheses that need to be substantiated with evidence from the individual, not as blueprints that apply to all women.

Specific Counseling Skills Necessary for Women Clients

The best resource for identifying the essential counseling skills and attitudes for working with women clients is *Principles Concerning the Counseling and Psychotherapy of Women* (Division 17, American Psychological Association, 1978). The following comments rely heavily on this document.

Much of scientific knowledge of human development, human personality, and dysfunction is based on male subjects and male gender role definitions. Historically, the tendency was for scholars to assume that what was true for males could be directly applied to females. When theories of female development were proposed, they tended to use male experience as the starting point and therefore often greatly distorted female development. Freudian theory, for example, has been especially criticized for this practice. The male bias in psychology and human development has meant that the unique aspects of women's experience have been ignored or misunderstood. Many training programs in counseling and human service have educated their graduates to assume that male definitions of health, personality, and dysfunction can be directly applied to all human behavior. Currently, training programs are improving their curricula to have a more inclusive scope, but many counselors still lack information about the male bias of traditional human development theories and about the newer theories that are more inclusive. Thus, the first step in learning new skills is education about female development, female experience, and the role of gender bias in defining health and dysfunction. Those who have not been exposed to courses in the psychology of women or gender issues in counseling should seek out such courses. Books that familiarize counselors with basic concepts are readily available. Those listed in this chapter's References represent an excellent starting point.

Second, because women's motivation to seek counseling is often related to experiences of discrimination, victimization, and prejudice, counselors need to have a repertoire of counseling interventions that includes strategies to help individuals deal with injustice. Models of dysfunction that assume that the whole problem is internal or intrapsychic ignore the experiences of women (and culturally diverse groups in the society). Feminist therapists use the term *empowering clients* and talk of the need to advocate on behalf of the client to address the injustices she is experiencing (Rosewater & Walker, 1985; Worell & Remer, 1992). They suggest that a major goal of therapy is to help the client use the power she does have to intervene against unjust treatment. The role of advocate is a controversial one, but the important point in this debate is the need for the counselor to acknowledge the social and political influences on the individual. Acknowledgment of social influences on human behavior is embraced by virtually all models of counseling and therapy, not just feminist ones. The recognition of social injustice as a barrier to effective functioning is especially important in light of two factors: the female client's tendency to blame herself for victimization experiences that are objectively not her responsibility at all; and her risk for continued victimization if her inappropriate guilt and poor self-esteem are not addressed.

Third, counselors need to avoid sexist language in all their professional work. Guidelines for inclusive language are printed in the *APA Publication Manual* (1983). Referring to all humans as "men" or adult females as "girls" ignores the existence of women and their right to full adult status. Although some have claimed that such changes are simply semantic and superficial, the more accepted view is that in an enterprise as dependent on language as counseling, the particular words chosen are indeed critically important to empathy, respect, and openness to change.

Fourth, counselors need to refrain from social and sexual dual relationships with female clients. The evidence indicates that many women are victims of unscrupulous therapists who use them to gratify their own needs. The culture's message that submissiveness is equivalent to mental health for adult women is one contributor to this vulnerability. A second contributing factor is women's history of victimization by others in their lives. Counselors must also avoid creating other forms of dependency in any female client such that the autonomy of the woman is jeopardized. Counselors who find themselves instructing women clients about how to live their lives or making their decisions for them have violated this principle.

Just as counselors ought not to condone racist or culturally insensitive behavior by colleagues, they should not accept or condone sexist attitudes and behaviors toward women. Instead, counselors have a responsibility to educate colleagues, supervisors, and students about sexism and gender bias. They also have a responsibility to work with institutions and other professions to change policies and practices that are inherently unfair. Not only should counselors help clients realize their own power to respond to unjust practices by individuals and institutions, they also have an affirmative responsibility to get involved in combating sexist practices in institutions where they have some affiliation. For example, a counselor in a school or community agency has a responsibility to be helpful in the development and enforcement of sexual harassment policies.

Gender Bias toward Men

Just as the culture carries messages about appropriate behavior for women and girls, it defines acceptable behavior for men. Sexist attitudes toward men in Euro-American culture include the belief that men must be unemotional, self-sufficient, aggressive, and competitive. The value of men is usually measured by their economic success and capacity to support women and children. Males who do not fit this rigid definition of acceptable male behavior may experience negative attitudes on the part of others and feelings of low self-esteem. Even those whose behavior does comply with the gender role stereotypes often find their lifestyle and interpersonal relationships unfulfilling and constraining. For traditionally acculturated males, seeking counseling can be seen as a sign of weakness and failure to be independent, unemotional, and self-sufficient. Counselors need to be sensitive to the way gender role stereotypes have affected their male clients and must recognize that self-exploration, affective awareness, and admission of interdependence with others may be difficult for many males. At the same time, counselors have also internalized cultural norms about acceptable male behavior and need to examine their own gender role stereotypes so that they do not inadvertently label any male behavior that is deviant from cultural norms as pathological. A study by Robertson and Fitzgerald (1990) indicates that the issue of gender bias in therapists and counselors is not limited to their attitudes and behaviors toward female clients. In this research, males who chose nontraditional occupations and family roles were significantly more likely to be labeled as pathological by the practicing counselors and therapists participating in the study.

The Special Experience of the Culturally Diverse Woman

Women from culturally diverse backgrounds are often victims of double discrimination. For example, girls get less teacher attention and reinforcement in math classes than their male counterparts, but when the ethnicity of female students is taken into account, African-American girls get the least attention and reinforcement of all math students. Thus, effective counseling requires that this double discrimination be taken into account in the definition of problems of a minority female client and in the therapeutic interventions designed to help her resolve those difficulties. Counselors must take into account the interplay between a woman's ethnicity and gender. For example, Saunders-Robinson (1991) argues that dealing with battering by a male partner can be especially difficult for an African-American woman because of her reluctance to expose her male partner to the risk of racist treatment by police and the courts. This reluctance is supported by the tendency of the African-American woman to see the discrimination and prejudice she does experience primarily as due to racism rather than sexism (Jordan, 1991). Similarly, when working with Latina women, counselors need to be sensitive to their clients' experience in the majority culture and the Latino culture's definitions of appropriate roles for males and females (i.e., the concepts of *machismo* and *marianismo*). Arrendondo (1991) presents an excellent discussion of this topic. In short, counselors who work with culturally diverse women need to be well informed about the culture of the client, including specific cultural gender role definitions, and about the role of racism and sexism in the difficulties the client presents in counseling. In addition, counselors must be patient with the pace of developing trust and should be prepared to work hard to demonstrate credibility and trustworthiness.

Case: Doris

Doris is a 24-year-old accountant working for a large firm in a Southern city. She came to the mental health center at the encouragement of her closest friend, who was very worried about Doris. Doris described feeling worthless and sad for several months and remembers that these feelings started all at once rather than gradually. However, she can't recall any specific event that triggered them. Along with those negative feelings have come periods of high anxiety and bouts of sleeplessness. She says she feels removed from other people and is having a hard time relating to anyone. Lately, she has taken many sick days from work and even visited her doctor because of these problems. Her physician also recommended counseling as a first step when Doris asked for tranquilizers. Doris says she has not had any suicidal thoughts but is getting discouraged by the fact that none of her usual coping strategies seem to be working. Typically, when she feels blue or jittery, she exercises more or distracts herself from the problems with friends or hobbies. Doris is a fourth child of an intact family who all live close to her. She has several close friends and has dated a lot since college but is not interested in a committed relationship now, or at any time in the near future. Doris does not think she ever wants to become a parent because of her dedication to her career.

Given what you already know about Doris, think for a few minutes about what additional information you would find necessary to help this client. If possible, discuss this case with a classmate, and write down the additional areas you would like to explore. Then, examine your rationale for those choices. Finally, suppose that Doris were a Donald. Does that change any of your thinking? Examine closely whether that affects your view of Doris's decision not to have children or her lack of interest in marriage. Again, talk with a partner if possible about the implications of gender. Did you take into account the possibility of an experience of victimization such as acquaintance rape or sexual harassment that could have triggered the depressed and anxious feelings? Even though she says she does not recall a trigger event, it is possible that one occurred but she is repressing it or simply did not label it as a victimization experience. Such a response is not uncommon among women and is perhaps made even more likely by Doris's usual degree of control over her life and future. Another possibility is that Doris may remember but feel ashamed about disclosing it to a stranger. The fairly sudden onset of these feelings in a female who otherwise functions well should be cause to explore that possibility along with other experiences of loss or disappointment.

SUMMARY

Because of population changes and increased access to education and mental health services, counselors today must be prepared to work with an increasingly multicultural clientele. For the counseling process to be successful with culturally diverse clients, counselors must explore their own cultural traditions and attitudes to eliminate cultural elitism and gain a true appreciation for the value of cultural diversity in society. Knowledge of other cultures is also important so that counselors can accurately define client problems and choose appropriate intervention strategies. Furthermore, counselors must understand how racism and cultural oppression have affected the lives of clients and how these factors may affect the counseling relationship. Specifically, culturally diverse clients faced with a counselor of a different background may be cautious, skeptical, or even openly mistrustful. In addition, counselors need specific skills to overcome these barriers and help the client solve the problems that brought him or her to seek help. These skills include bilingualism, a broad repertoire of counseling strategies, appropriate use of assessment tools, a list of referral and consultation sources for multicultural clients, and the ability to work effectively with support persons in the community.

Women and girls are also considered a population for which effective counseling requires that counselors have special skills. Women have experienced gender role stereotyping that may affect the kinds of concerns they bring to counseling and their risk for victimization. Counselors need to understand the effects of sexism on female clients' development, the society's current values and norms for human behavior, and how sexism may interfere with the counseling process. Specifically, counselors need education in inclusive theories of human development and skills for helping girls and women who are currently experiencing victimization and discrimination.

REFERENCES

American Association of University Women. (1992). *How schools shortchange girls.* Washington, DC: Author.

American Psychological Association. (1983). *Publication manual* (3rd ed.). Washington, DC: Author.

American Psychological Association. (1990). *Women and depression.* Washington, DC: Author.

Arrendondo, P. (1991). Counseling Latinas. In C. C. Lee and B. L. Richardson (Eds.), *Multicultural issues in counseling: New approaches to diversity* (pp. 143–156). Washington, DC: American Association for Counseling and Development.

Atkinson, D. R., Morten, G., & Sue, D. W. (1993). *Counseling American minorities: A cross-cultural perspective.* Dubuque, IA: Wm. C. Brown.

Belenky, M. F., Clinchy, B. M., Goldberger, N. R., & Tarule, J. M. (1986). *Women's ways of knowing: The development of self, value and mind.* New York: Basic Books.

Broverman, I. K., Broverman, D. M., Clarkson, F. E., Rosenkrantz, P. S., & Vogel, S. R. (1970). Sex role stereotypes and clinical judgments of mental health. *Journal of Consulting Psychology, 34,* 1–7.

Division 17, American Psychological Association. (1978). *The Division 17 principles concerning counseling/psychotherapy of women.* Washington, DC: Author.

Estes, C. P. (1992). *Women who run with the wolves.* New York: Ballantine.

Gilligan, C. (1982). *In a different voice: Psychological theory and women's development.* Cambridge, MA: Harvard University Press.

Herr, E. L. (1989). *Counseling in a dynamic society: Opportunities and challenges.* Alexandria, VA: American Association for Counseling and Development.

Ivey, A., & Authier, J. (1978). *Microcounseling: Innovations in interviewing training.* Springfield, IL: Charles C. Thomas.

Jordan, J. M. (1991). Counseling African-American women: "Sister-friends." In C. C. Lee and B. L. Richardson (Eds.), *Multicultural issues in counseling: New approaches to diversity* (pp. 51–63). Alexandria, VA: American Association for Counseling and Development.

Lee, C. C., & Richardson, B. L. (Eds.). (1991). *Multicultural issues in counseling: New approaches to diversity.* Alexandria, VA: American Association for Counseling and Development.

Pedersen, P. B. (1988). *A handbook for developing multicultural awareness.* Alexandria, VA: American Association for Counseling and Development.

Pedersen, P. B. (Ed.). (1991). *Journal of Counseling and Development, 70* (entire issue).

Robertson, J., & Fitzgerald, L. F. (1990). The (mis)treatment of men: Effects of client gender role and life style on diagnosis and attribution of pathology. *Journal of Counseling Psychology, 37,* 3–9.

Rosewater, L. B., & Walker, L. E. (Eds.). (1985). *Handbook of feminist therapy: Women's issues in psychotherapy.* New York: Springer.

Ryan, W. (1971). *Blaming the victim.* New York: Pantheon.

Saunders-Robinson, M. A. (1991). Battered women: An African-American perspective. *ABNF Journal, 2,* 81–84.

Sue, D. W. (1978). Eliminating cultural oppression in counseling: Toward a general theory. *Journal of Counseling Psychology, 25,* 419–428.

Sue, D. W., Arrendondo, P., & McDavis, R. J. (1992). Multicultural counseling competencies and standards: A call to the profession. *Journal of Counseling and Development, 70,* 477–486.

Sue, D. W., & Sue, D. (1990). *Counseling the culturally different: Theory and practice* (2nd ed.). New York: Wiley.

Tannen, D. (1990). *You just don't understand: Women and men in conversation.* New York: Morrow.

Worell, J., & Remer, P. (1992). *Feminist perspective in therapy.* New York: Wiley.

Wrenn, C. G. (1962). The culturally encapsulated counselor. *Harvard Educational Review, 32,* 444–449.

CHAPTER 10

WORKING WITH RELUCTANT CLIENTS

What has been said in the preceding chapters applies most readily to voluntary, or self-referred, clients who are willing to initiate discussion on an issue of concern. Starting the counseling process with such clients is relatively easy—the counselor simply begins with a statement that provides the client with an opportunity to share information. "How can I help?" is one example. Responses such as summaries or reflections of feelings and beliefs, leads such as "Tell me more about . . . ," and "I" messages help to facilitate further exploration. Readiness to participate openly in the counseling process ranges all the way from the client who seeks counseling to the client who is referred by a third party or encounters the counselor as part of a school guidance program to the reluctant client.

WHO IS THE RELUCTANT CLIENT?

The reluctant client is any person who is unusually mistrustful of the counselor. Such clients "feel attacked or accused or unsupported" by the helper and "are likely to put on the brakes" (Pipes & Davenport, 1990, p. 171). Many reluctant clients, if given a choice, would choose not to be in the presence of a counselor and would prefer not to talk about themselves. Many are not self-referred but are instead required or urged to engage in counseling by court action, school officials, spouses, or parents. Some self-referred clients also have a lot of ambivalence about seeking help from others and may be mistrustful of the counselor's motives even though they have tentatively decided to give counseling a try. In certain settings, "these [reluctant] clients represent the major proportion of the counselor's caseload" (Ritchie, 1986, p. 516).

Dyer and Vriend (1973) stated that a large proportion of students who visit school counselors are not there entirely voluntarily. These include students whom parents, the counselor, or other school staff have identified as academically or behaviorally deficient. Referrals from teachers and parents are often made with the implicit understanding that the counselor will straighten out the client's "silly" thinking and show him or her a proper course of action.

In clinical settings, many clients appear for counseling at the behest of significant others, including spouses, lovers, parents, or children. In marriage counseling and counseling about problems encountered in intimate relationships, one party in a relationship often brings his or her partner to counseling. Parents frequently bring their children to counseling, and, increasingly, adolescent and adult children are responsible for bringing their parents to counseling. Agendas including substance abuse and family violence, among others, result in family members insisting that other family members attend. Finally, the justice system may make counseling a requirement for juveniles or adults who have violated various laws. Some clients referred by the justice system are incarcerated, and others are required to seek counseling as a condition of remaining out of confinement while working toward rehabilitation.

Some adults, particularly males, consider that it is a sign of weakness to need help in resolving life's problems (Mintz & O'Neil, 1990; Scher, 1981; Scher, Stevens, Good, & Eichenfield, 1987). Even if such persons initiate counseling, they may resent having to do so and behave as reluctant clients. Persons belonging to certain ethnic minorities (e.g., Asians) may be socialized to be very private about their concerns and will have difficulty with disclosure. Other minority clients who feel they have been wronged by the majority society may bring anger about racism to the counseling situation. (See Chapter 9 for a more complete discussion of working with the reluctance of minority clients.)

When a counselor meets a reluctant client for the first time, what occurs often seems irrational to the counselor, as it would to an impartial observer looking in on the session. The counselor offers an accepting welcome, attends caringly to the client's communications, shows genuine interest in the client and the difficulties he or she is experiencing, and expresses empathy for his or her situation. In return, the client is often initially sullen, rude, silent, belligerent, or overly compliant ("Just tell me what you want from me"). The counselor may feel a sense of rejection and even receive verbal abuse from the client, whom the counselor does not yet know and to whom he or she has done nothing offensive. The natural human response to such attack is to be angry, and indeed the counselor may experience a flash of anger toward such a client.

UNDERSTANDING THE CLIENT'S RELUCTANCE

In this chapter, reluctance is seen primarily as a function of the phenomenon of transference. To understand why the counselor is treated like an enemy when he or she acts more like a friend, we look to the client's expectation of how he or she will be treated. Clients who have a history of conflict with parents, teachers, employers, and other persons in positions of authority tend to expect conflict in this new relationship with a person whom they probably perceive as another authority figure. They exhibit the attitudes and behaviors that they have learned and practiced in their earlier dysfunctional relationships with people who were in positions to exert influence over their lives. "It is as if the client is so certain that what he or she is experiencing is unacceptable that the therapist is presumed to be unaccepting and critical, all evidence to the contrary" (Pipes & Davenport, 1990, p. 173).

Transference (Brammer, Shostrum, & Abrego. 1989, p. 200) is "repetition of past conflicts with significant others such that feelings, behaviors, and attitudes belonging in those early relationships are 'transferred' or projected onto the therapist." According to transference theory, a person may build a pattern of negative and self-defeating response to authority over an extended period of time. The pattern usually begins with parent-child relationships in which the child perceives that the parents make arbitrary demands and make their love conditional upon the child doing their bidding. The child experiences hurt, deprivation, frustration, and eventually anger because his or her behaviors are perceived by the parents as wanting and love from them is not dependable.

As the child expands his or her world of social contact, teachers are the next authorities with direct influence over his or her life. If the parent-child relationship does not improve and the child remains insecure about whether he or she is competent and loved, any demand a teacher may make is a further threat, and some of the resentment originally held for the parent is transferred to the teacher. Even reasonable expectations of teachers may prompt such transference if the situation with parents is strongly negative, and the situation may get worse if the teacher makes unreasonable demands or reacts hostilely to the resentment the child holds. Sometimes this kind of sequence is repeated as the young person meets more new authority figures, and so by adolescence he or she perceives all adults as controlling and uncaring. A pattern of self-fulfilling prophecy develops. Anticipating negative responses from adults, the young person behaves in a way that tends to elicit such responses. If these authority problems are not resolved during adolescence, they are likely to manifest themselves through the adult years in situations involving family life, employment, and other socially defined roles. Even though a counselor may not perceive himself or herself as an authority figure, to a client with an authority problem, the counselor is one more person trying to get him or her to change. However kindly the counselor may proceed, the client is likely to employ the strategies that he or she usually uses with authorities.

When a person perceives authority figures as uncaring and demanding, two kinds of reaction are typical. The first and more common is a hostile stance that tests each new authority figure and puts him or her at a distance. The other is excessive compliance, perhaps initially based on the misguided hope that doing exactly what is requested will win elusive acceptance. But compliant personalities lose control of their own lives, and when things do not work out, often respond, "I only did what you told me to do." Either reaction is detrimental to counseling because the former militates against the quick formation of a trusting relationship (therapeutic alliance) and the latter leads to dependency. Dependent clients look to the counselor for answers and do little productive work on their concerns.

RELUCTANCE AND RESISTANCE

Resistance is an unconscious process whose purpose is to protect the client from having to explore and claim feelings and motivations that have roots in his or her past. "Resistance includes everything in the words and behavior of the client that prevents access to unconscious material" (Ivey, Ivey, & Simek-Morgan, 1993, p. 206), and thus resistance opposes the purposes of counseling (Brammer, Shostrum, & Abrego, 1989). It is an intrapsychic process that is experienced by all clients (not just reluctant ones). A client who has urges toward growth and completeness nevertheless still fears the pain of recalling traumatic material (Pipes & Davenport, 1990) and resists abandoning the certain present—with all its limitations—in favor of an unknown future. A reluctant client such as we have discussed is quite aware that he or she does not have much trust in the counseling process and would prefer not to participate. To the degree that this reluctance emerges from transference of feelings, it has origins that may not be understood by the client. But reluctance has an interpersonal as well as an intrapsychic component; the client's feelings and attitudes about counseling and the counselor are often expressed very directly.

In a sense, a reluctant client acts out a part of his or her issues with the counselor, but a resistant client's struggle occurs among competing motives within himself or herself. Naturally, it can be expected that a reluctant client will also be resistant in other ways, since resistance has been described as a universal experience in counseling.

THE COUNSELOR'S EMOTIONS TOWARD THE RELUCTANT CLIENT

As Dyer and Vriend (1975) describe, counselors often experience relationship-blocking emotions when faced with a reluctant client. The counselor has a reality-based concern about whether counseling will succeed and the noxious experience of feeling rejected by another person. On the other hand, the counselor may experience countertransference reactions to the client that call forth his or her tendency to respond in authoritarian (controlling parent) or dependency-building (solicitous parent) ways. Such countertransference reactions are based upon the counselor's unresolved feelings about the nurturing process, and as with transference, the roots of the feelings are largely beyond the counselor's awareness. The sources of the counselor's emotions when working with reluctant clients and some strategies for managing those emotions are developed in this section.

It is natural for a counselor to experience some anxiety when faced with a reluctant client. The client can be a threat to the counselor's sense of professional competence and is potentially capable of blocking the counselor's goal of being a successful helper. Opening communication, building trust, stimulating exploration, developing insights, and changing behavior are all made more difficult. If the counselor's anxiety about the possibility that counseling might fail is not managed successfully, his or her ability to respond sensitively and insightfully may be impaired. Counselor anxiety can lead to frustration and even anger at the client who is obstructing successful helping. Ironically, this amounts to blaming the client for his or her resistance rather than accepting it as a natural part of counseling that must be worked through.

As the encounter with a reluctant client takes shape, the counselor should monitor his or her self-talk about the client. This is one possible perspective: "This client has a lot of nerve. What right has he or she to treat me so disrespectfully? I don't see how I can be of much help with his or her attitude as it is." The focus here is on the counselor's offended feelings and fear of failure. The client is blamed for these feelings, and counseling will be difficult. On the other hand, consider this counselor self-talk: "This client has barriers that are very strong. I will have to work hard to earn his or her trust and to remember that his or her behavior makes sense when viewed from his or her experience. I can't take his or her attacks personally if I want to be able to help." The focus is on the client and demonstrates an emotionally neutral acceptance of the client's behavior.

Counselor feelings that are countertransference reactions are hardest of all to manage, for they stem from the counselor's unconscious predispositions about authority relationships. Fundamentally, the problem begins when the counselor at some level decides that the client is incapable of handling or learning to handle his or her own life situation, and so assumes an authority role with respect to the client. Having made this decision, the counselor responds as a controlling parent might, telling the client what he or she should do. With a dependent client, the counselor's response may be to try to do everything *for* the client rather than urging self-direction. Either way, the counselor loses sight of the fact that the client cannot grow in his or her ability to handle life situations unless he or she is an active participant in decisions that are made. Even very skilled counselors are sometimes overcome by their needs to be parental, especially in interactions with children and youths.

Perhaps the greatest problem with countertransference is that it exacerbates the client's transference behaviors. With a hostile client, a counselor who acts in an authoritarian way fulfills the client's expectation that the counselor is an authority figure who is trying to manipulate. The counselor can no longer offer the unique relationship (caring, genuine, and unconditional) that characterizes the helping process. When this set of events unfolds, we say that the counselor has fed the transference and reinforced the client's self-defeating views that all people who are "part of the system" are controlling and insensitive. If the conflict is not resolved, such a client will be even more reluctant to seek counseling again.

If a client has tendencies to be dependent on authority figures, a counselor acting as a "solicitous parent" will increase the client's tendency to avoid responsibility. There are many examples of clients who hound their counselors for advice about the smallest move, "hang out" in the guidance office, or call counselors at home about inconsequential matters. This is the result of the counselor feeding the client's dependency needs, perhaps to satisfy his or her own countertransference needs to be a caretaker.

Perhaps the most important concern about the impact of countertransference is the fact that the client usually does not simply stay the same. The client often becomes more dysfunctional and less confident that the counseling process can be helpful. Probably the most effective way of coping with serious countertransference problems is through consultation with or supervision by another counselor. As a general rule, any time a counselor thinks "If I could only control that crazy behavior" or "If I could only take that poor dear home," that is the time to talk with a colleague. By discussing his or her feelings with another counselor, a counselor in this kind of situation can come to a better understanding of his or her own needs and restore his or her ability to place the client's needs first. Some counselors-in-training may find that their need to assume responsibilities that should be the client's is strong. In such cases, the counselor himself or herself may need counseling to resolve countertransference needs.

Reluctant clients create stress for *all* counselors. The counselor's response to this stress is influenced by the persistence and degree of dysfunction of the client's behavior as well as countertransference based on the counselor's unresolved needs. In many instances, counselors can improve their work with reluctant clients through skill development or through self-reinforcing talk. In other instances, consultation or counseling with a more objective third party (another counselor) is needed.

WHY WORK WITH RELUCTANT CLIENTS?

Whether or not counselors should be expected to work with reluctant clients is an important issue. Some say counselors should work only with those who want a counselor's help. Others argue that, as professionals, counselors should be expected to work with anyone who stands to benefit from their services.

Counselors in schools are expected to work with students whose academic performance is unsatisfactory, whose behavior is disruptive, or who seem to have no sense of personal direction or commitment to the future, whether or not those students seek help. These behavior patterns are often important indicators of poor self-esteem. A counselor working within the framework of a developmental guidance program will extend counseling services to such students with the goals of encouraging them to identify and resolve underlying problems, to identify motivating forces, and to come closer to fulfilling their human potential.

Counselors who work in the justice system know when they sign an employment contract that many of their clients will be reluctant about counseling. The decision to work with such clients is based on the belief that persons who commit crimes can learn to fulfill their needs in ways that don't violate the rights of others and that avoid future incarceration. To the degree that this occurs, both the individual and society benefit.

Many clients in substance abuse counseling have lost control of their lives to the substance (alcohol or other drugs) they are using. Their motivation is to maintain their habit even though they may recognize that they are losing their grip on work and family. It is often the family and/or the employer that insists that counseling be sought—for the client's own welfare and that of the family.

Finally, there are the clients who are referred or brought to counseling by family or friends who are concerned about the client's mood, behaviors, decision making, or performance. Many times such clients do not see themselves as needing help—yet in many instances an objective third party will see patterns of self-defeating and ineffective behavior.

There are some important assumptions underlying a counselor's decision to work with a reluctant client. One is that there is a moral and ethical obligation to help others live more fully if one has the capacity to do so, whether or not those others seek help. This assumption is supported by the knowledge that an individual who has not experienced counseling cannot know its value.

A second assumption is that the counselor has the capacity to recognize ineffective behavior or unhappiness that may be amenable to alteration through counseling. The counselor's model of the mentally healthy individual forms a standard by which to evaluate client behaviors and emotions and to identify unhealthy or undesirable patterns. What may appear as unhealthy on the surface may turn out to be reasonable coping upon further understanding of the client and his or her situation.

Another assumption is that early intervention for dysfunctional behavior is likely to lead to more rapid and more complete resolution of the problem. This may serve as justification for a counselor to begin working with a client before the client recognizes the need for help. Examples include intervention with academic problems before skills deficits make it difficult to catch up and intervention with an alcoholic prior to the time the person "hits rock bottom." Counseling is used to help prevent problems from becoming more serious or from extending their debilitating influence from one aspect of the client's life (e.g., work) to another (e.g., marriage).

When a counselor decides to work with a reluctant client, he or she concludes that it would be better to try to influence the client than to allow things to go on as they are going. To this degree, the counselor assumes some control over what happens. There is a judgment made that the client will benefit, and there is also a risk that the counselor will be seen by the client to be like others who have judged him or her to be insufficient. The counselor's effectiveness lies in his or her ability to communicate caring and trust, to avoid a moralizing posture, and to avoid using coercive tactics to get the client to conform to someone else's norms. The reluctant client, like any other, deserves the opportunity to design his or her own solutions to life's difficulties.

WORKING WITH THE CLIENT'S RELUCTANCE

In earlier sections of this chapter, we have defined reluctance and the related concept of resistance. Both processes oppose the purposes of counseling when manifested by the client, even though the client is usually aware of reluctance and not aware of resistance. Fear of trusting and fear of changing underlie both processes. Since all clients experience some resistance in counseling, it may be assumed that a client who is actively reluctant is experiencing resistance as well. Authors who discuss reluctance and resistance as separate issues often conclude that treatment is similar whichever process is occurring (Egan, 1990; Pipes & Davenport, 1990). It is essential that the counselor join the client in exploring what is making it difficult for the client to trust the counselor and to begin work on his or her problem (Pipes & Davenport, 1990). The counselor must avoid an adversarial relationship. As Cormier and Hackney (1993) so clearly state, the counselor should never fight with the client.

The first principle is to start where the client is. If the counselor remembers that the reluctant client is fearful of what may happen in counseling and angry at having been coerced into participation, it is much easier for the counselor to avoid reacting to client affect as though it were intentionally directed at himself or herself. Quite literally, to the extent that the counselor can truly join the client's struggle about being in counseling and possibly needing to change and can keep the focus on the client's experience, little of the counselor's attention will be directed to feelings associated with being manipulated by the client in hostile or dependent ways. Both the counselor's conscious emotional responses and his or her countertransference will remain under control.

Accepting a reluctant client involves accepting the client's reluctance as a part of the agenda of counseling. One might begin with "Help me understand what has been happening to you." The counselor accepts the perspective that the client "does not want to talk with me right now, and that is his or her right." But at the same time, the counselor persists with leads that show genuine caring and involvement. Eventually, the client may begin to observe that the counselor is not behaving like authority figures typically do when he or she is resistant. At that point, the client has to find other ways to interact that make sense in the new environment, and the beginnings of trust are established.

Pipes and Davenport (1990) recommend a three-step process of helping the client to become aware of his or her reluctant behavior. The first step is simply to observe the behavior, for example: "You seem to be very angry." The second step is to place the feeling into some kind of context: "You are angry because you were required to come here and you don't see how I can help." The third step is to openly invite the client to discuss his or her feeling: "Would you care to explore what it is that upsets you so much?" In actual experience, we have found that many clients respond so fully to the first lead that the subsequent ones become unnecessary.

For the client with an authority problem, it may take a number of sessions to build an understanding of whom he or she is angry at and what he or she may be able to do about it. The client may also need to be told when he or she is projecting that anger onto the counselor or other persons who have not earned it—though such responses are confrontative and thus are most effective once the client has begun to trust the counselor and the process. Through such discussion, the client can learn to discriminate between people who have perhaps committed wrongs against him or her and those who have simply been assigned "bad guy" status. Then the process of working through feelings toward the original objects of the anger (usually parents or siblings) may proceed. The client can be helped to develop an understanding that the past—no matter how deplorable—need not continue to affect the present. With a client whose resistance relates back to family relationships of long standing, change will be slow, and the client will require much support before trying new behaviors. The client will experience many failures with his or her new behaviors and will have intense feelings about the failure. The path to more positive feelings about self will be slow.

Remember that all reluctant clients are a challenge to the counselor and that the chances of success are smaller than with self-referred clients. A counselor who undertakes working with reluctant clients must have the commitment to stick with the process and not become just another person who eventually gives up. Giving up will be seen as just one more rejection that further builds the client's view that authorities can't be trusted. On the other hand, many reluctant clients leave counseling of their own volition before the process reaches fruition. It is important that the counselor recognize this likelihood and acknowledge such failures as the inevitable outcome of reaching out to those who are hard to help. It is important to take satisfaction from successes rather than castigating oneself for failures.

WORKING WITH THE PERSON MAKING THE REFERRAL

Since a large number of reluctant clients are referred to counseling by a third party, we conclude this chapter with some important principles for managing contacts with these referring parties. Most often, the referring party has a relationship of some duration with the client and is in a position to provide information that will be useful in the counseling. In some cases, the reason for referral is disruption in the relationship between the client and the referring party. At other times, the referring party has observed client behaviors that have caused concern for the client's welfare.

When a third party makes a referral, the counselor tries to learn as much as possible about why the referral is being made and should inform the referring party of the nature and scope of services that may be made available. Discussion of confidentiality may also avoid later misunderstandings.

Often referral sources are vague about their reasons for wanting a counselor to work with an individual. Just as it is necessary for counselors to help clients clarify their problems, it is also important to assist persons making referrals to state their observations in clear and specific terms and to distinguish between their observations and inferences. It is not sufficient for a teacher referring a student to say: "Lucinda has changed a lot during the last marking period. She used to be carefree, but now she seems to be very worried about something. I think you should see her." The counselor should help the teacher describe the behaviors and circumstances that led to the inference that Lucinda is worried. "Can you tell me about some of your observations of Lucinda's behavior that have given you the impression she is worried?" is an example of a counselor response that can facilitate such clarification.

The counselor has many options in referral situations. In any setting, there are some boundaries on the types of clients who may be accepted. Certain agencies will accept most clients who need mental health counseling but may not have the resources to manage severely disturbed clients who pose a danger to themselves or others. Other agencies specialize in career counseling, marriage counseling, or some other specialty and accept only clients who fit their identified niche. School counselors work with students with a wide variety of learning, decision-making, and personal problems but generally refer those whose emotional problems lie outside the range of normal functioning. Similarly, school counselors often provide consultation to parents but usually refer families that need extended family counseling. Instances of child abuse are usually referred as well.

If the referring party describes a client whose concerns are obviously outside the scope of service of a counselor, that counselor may suggest another resource without ever meeting the client. More frequently, in general mental health facilities as well as school counseling offices, an initial meeting is scheduled with the client (often referred to as an intake interview), and a determination is made at that time as to whether the client could benefit from the services available at that location or should be encouraged to go elsewhere. The referring party should always be assured that the counselor will make every effort to ensure that needed help is provided but that he or she may not be the service provider.

In many instances, the identity of the referring party will be obvious to the client. In the criminal justice system, it is quite clear when a judge makes counseling a condition of probation, for example. If a parent brings a young person into a counselor's office, there is little doubt in anyone's mind who it is who thinks that help is needed. In other cases, though, the referring individual may not have discussed the referral with the client. This frequently happens in schools: a parent or teacher will approach the counselor without talking with the young person about it. A foreman may contact a counselor in an employee assistance office to report that a worker seems to be experiencing a problem with alcohol or simply that his or her work habits have changed.

It is usually desirable for the counselor to tell the client who has been referred why the session has been arranged and who was responsible for initiating it. Such disclosure places the cards on the table and contributes to a genuine interchange. The referring party should be told about the disclosure during the referral discussion and should participate in deciding how much the client should be told and how.

The referring party and the counselor must also agree in advance about what kind of feedback the counselor will offer after counseling has begun. This can be a difficult dilemma for the counselor. Referring persons, especially parents and spouses, generally want some feedback, yet the counselor wants to offer the client an opportunity for private communication. Some counselors handle this dilemma by simply explaining the dilemma to the referring person: "I can well understand your desire for feedback. You care for Lucinda and want to know that something is being done to help her. But it is also important for me to help Lucinda develop a sense of trust. This means I must let her know that I will keep private whatever she shares with me unless she makes reference to hurting herself or others. Of course, she may agree to the sharing of certain information with you (or releasing information to others in a position to be helpful), but I hope you understand that I won't be able to share everything that we discuss." Sometimes referring parties are especially concerned about what the prospective client may say about them. The counselor can indicate that he or she will try to find ways of working on the relationship between the referring person and the client if that proves to be an important issue. Nevertheless, it is still necessary that the referring person understand that the client will have choices about what information is shared once the counseling begins.

Case: Eddie

Eddie, a 12-year-old seventh grader was referred to the counselor by his teacher (in an eight-grade elementary school with self-contained classrooms). Eddie was doing passing work, but his behavior suggested that something was wrong. He rarely spoke to other children in his class and spent his free time brooding by himself. He occasionally showed bottled-up anger by responding violently with little provocation, striking out physically at nearby people and objects. These "tantrums" were regarded as somewhat dangerous by the referring teacher. Furthermore, when they occurred, they seemed to increase Eddie's isolation even more. Eddie would respond to verbal questions from the teacher in class if he knew the answers to the questions, but the teacher had been unsuccessful in establishing any conversation with Eddie on an informal level outside of class.

The counselor in this case was in his mid-twenties and exuded caring as he worked with young people. He dressed informally and behaved in a fashion that allowed most seventh-grade boys to see him as a "cool" role model rather than an authority figure.

Initial Counseling Session

COUNSELOR: Hello, Eddie. How's it going?

EDDIE: Okay.

COUNSELOR: I imagine you're wondering why I wanted to see you. Your teacher asked me to talk to you. She seems to think you have some things on your mind that you'd like to talk to someone about.

EDDIE: Well, I don't.

COUNSELOR: Everything's fine with you, then.

EDDIE: Yeah. *(The counselor allows a fairly long silence to occur before continuing.)*

COUNSELOR: Your teacher says that you hardly ever talk to anyone and that you seem sad or worried or something. Do you think that describes how you are in her room?

EDDIE: Ain't got nothing to say.

COUNSELOR: So it is true that you don't buddy around much with the kids in your class?

EDDIE: They are stupid.

COUNSELOR: It must seem kind of lonely not to have anyone to talk to or do things with.

EDDIE: *(Silence—but visible emotion in the form of tears welling in his eyes.)*

COUNSELOR: Well, maybe you'd like to think about whether you can tell me how it is for you at a later time. Would you like to shoot a few baskets with me for the rest of our time today?

The counselor and Eddie adjourned to the playground, where they spent some time playing basketball. By mutual consent, the conversation was restricted to short comments about the play.

Subsequent Events

Two days after the counselor's initial meeting with Eddie, the boy became involved in another classroom incident that resulted in a conference with the teacher, the school principal, the counselor, and Eddie's mother. The mother described how difficult it was for her to "manage" Eddie by herself. She expressed the opinion that a boy needs a father to keep him under control and that she did not know what to do with him.

Since the problem seemed to indicate personal adjustment and development issues, the principal and the teacher left the counselor and Eddie's mother alone to consider what might be done. There was a long history of family violence, and both the mother and Eddie (an only child) had suffered physical abuse. The mother revealed that Eddie's father was in prison, having been convicted of second-degree murder. After the publicity of the arrest and conviction, the mother had moved with Eddie to a new town, where they and their history were not known. The incidents of Eddie's violent behavior at school occurred when other students tried to get close, tried to learn more about him or his family, or made any comments that he construed to reflect on his parents.

The counselor continued to meet with Eddie, who was told by his mother that the counselor knew about his father. At first, the sessions were still fairly difficult, with Eddie sharing very little of his thoughts or feelings. He did eventually show signs that he liked the attention from the counselor and enjoyed some of the activities they did together. The first indication that counseling was having an effect occurred after about a month, when the mother called to say how cooperative Eddie had become at home and that the evening before he had kissed her for the first time since his father had been arrested. Through his relationship with the counselor, Eddie slowly recovered his ability to trust and love other people, his interaction with other students improved, and he progressed toward a more satisfying lifestyle. (He never did discuss his feelings about his father with the counselor, and this could be unfinished business that may create further difficulties later.) Eddie gained from counseling in spite of some reluctance that persisted throughout the six-month period during which he worked with the counselor.

Discussion

Eddie's reluctance to work with the counselor was an extension of his reluctance to interact with other people. He had a history that was too painful to discuss even after the counselor was informed about it by the boy's mother. Eddie chose to keep his feelings about his father private, even through extended contact with the counselor. Therapeutic movement nevertheless occurred, presumably as a result of a caring and respectful relationship with the male counselor, who served as a person to be emulated. The fact that the counselor could accept and care for him, even knowing the dark secret of his father's crime, was in itself supportive. Counseling conversations eventually included content about day-to-day relationships as they began to build. Future-oriented discussion about education and career planning also took place.

Eddie's reluctance to share his world with others is easy to understand. Clearly, it would have been inappropriate and counterproductive for the counselor to respond in anger or frustration to Eddie's initial rejecting behavior. Through patience, persistence, and understanding, the counselor was eventually able to experience a measure of success in helping Eddie to lead a fuller, more satisfying life.

Questions for Further Thought

1. To what extent do you believe that transference may have been a factor in Eddie's reluctance to become involved with the counselor?
2. How important do you believe it was that Eddie and the counselor never openly discussed Eddie's feelings about his father?
3. Can you account for Eddie's reaction when the counselor observed that he seemed lonely? How lonely was he? What do you think about the counselor's decision to disengage from the counseling session at that point?
4. In the case study, it was stated that the counselor had a personal charisma that made him popular with young people. How important do you think that was? Your own personal characteristics are different in some ways from the counselor's. If you were Eddie's counselor, how might you vary the approach to take advantage of your own personality in building the relationship?

SUMMARY

A certain proportion of clients seen by any counselor will be reluctant clients—persons who would not seek counseling if the choice were left to them. They are requested or required to see a counselor by spouse, parents, teachers, or the legal system. They express reluctance by failing to participate fully in the counseling process and sometimes are hostile, discourteous, unpleasant, or dependent in their relationship with the counselor. In addition to the conscious interpersonal element that we have referred to as reluctance, these individuals experience resistance, or intrapsychic fears of change that oppose the work of counseling.

A client may be reluctant to participate in counseling because he or she feels that there is little to be gained by sharing private thoughts with another person. Perhaps he or she has been hurt in the past when trying to discuss personal thoughts and feelings. Most of the defensiveness centers on previous and current relationships with authority figures. Anger, frustration, alienation, or dependence that have been experienced in relationships with authority figures, usually parents, are transferred to the counselor. The counselor, who has offered help and kindness, is likely to experience feelings of rejection when faced with a reluctant client. Furthermore, the counselor may feel anxious about the possibility that counseling may fail. Nevertheless, the commitment to work with reluctant clients is important because it constitutes a reaching out to individuals who have usually built barriers to relating with others and who need help in establishing better contact with others.

The counselor can manage his or her own feelings of rejection by focusing on the client. Instead of personalizing the rejection, the counselor must see it as the client's present need and begin to work where the client is—with his or her reluctance to become involved. Having established such an initial approach, the counselor takes special care in using the relationship-building skills of the first stage of counseling. Consultation or supervision to help the counselor manage countertransference is often helpful. Many initially reluctant clients will eventually respond to the counselor's caring and genuineness and begin to trust him or her. To the degree that such a change occurs, the road to other kinds of change is opened. However, the success rate with reluctant clients is lower than with self-referred clients.

Working with the referring party is another delicate aspect of working with reluctant clients. The referring person should be helped to understand the nature of counseling and the importance of confidentiality. The degree to which the referring person is involved after counseling begins is dependent upon where he or she fits in the life of the client and the client's need and readiness for that person to be a part of the change process.

REFERENCES

Brammer, L. M., Shostrum, E. L., & Abrego, P. J. (1989). *Therapeutic psychology* (5th ed.). Englewood Cliffs, NJ: Prentice-Hall.

Cormier, L. S., & Hackney, H. (1993). *The professional counselor.* Boston: Allyn and Bacon.

Dyer, W. W., & Vriend, J. (1975). Counseling the reluctant client. *Journal of Counseling Psychology, 20,* 240–246.

Egan, G. (1990). *The skilled helper: A systematic approach to effective helping* (4th ed.). Pacific Grove, CA: Brooks/Cole.

Ivey, A. E., Ivey, M. B., & Simek-Morgan, L. (1993). *Counseling and psychotherapy: A multicultural perspective.* Boston: Allyn and Bacon.

Mintz, L. B., & O'Neil, J. M. (1990). Gender roles, sex, and the process of psychotherapy: Many questions and few answers. *Journal of Counseling and Development, 68,* 381–387.

Pipes, R. B., & Davenport, D. S. (1990). *Introduction to psychotherapy: Common clinical wisdom.* Englewood Cliffs, NJ: Prentice-Hall.

Ritchie, M. H. (1986). Counseling the involuntary client. *Journal of Counseling and Development, 64,* 516–518.

Scher, M. (1981). Men in hiding: A challenge for the counselor. *Personnel and Guidance Journal, 60,* 199–204.

Scher, M., Stevens, M., Good, G. E., & Eichenfield, G. (1987). *The handbook of counseling and psychotherapy with men.* Newbury Park, CA: Sage.

CHAPTER 11

WORKING WITH CLIENTS IN CRISIS

The counseling process, as described in the first seven chapters of this book, provides opportunities for clients to ventilate emotions, to examine the factors that have led to unwanted emotions and unsatisfying behaviors, to explore alternative plans of action, and to implement and test selected plans. The goals include the resolution of troublesome emotional experiences, the discovery of previously unexplored aspects of self, the planning of new courses of action, and a new integration of understandings about self in the environment. The client experiences growth in a range of interpersonal and instrumental functions of life. The process is deliberative and may require many sessions over an extended period of time.

When a client telephones or appears at the counselor's office in a state of crisis, priorities must be shifted to helping with the immediate crisis so that the client can regain the ability to manage the tasks of daily living as quickly as possible. Resolution of the crisis is sometimes all the help a client needs; in other instances, the client may have long-standing concerns that may become the content of further counseling once the crisis is handled. Aguilera (1990) describes crisis intervention as offering "the immediate help that a person in crisis needs to reestablish equilibrium" (p. 1).

DEFINITION OF CRISIS

We say that people are in a state of crisis when they perceive "an event or situation as an intolerable difficulty that exceeds [their] resources and coping mechanisms" (Gilliland & James, 1993, p. 3). Solutions that have worked before are no longer sufficient. The difficulty involves one or more life goals that the person fears are being blocked. As tension and anxiety over the inability to resolve the problem increase, the person becomes less and less able to find a solution. He or she feels helpless, upset, shamed, guilty, and unable to act on his or her own to reach a resolution.

Figure 11.1 Paradigm: The Effect of Balancing Factors in a Stressful Event

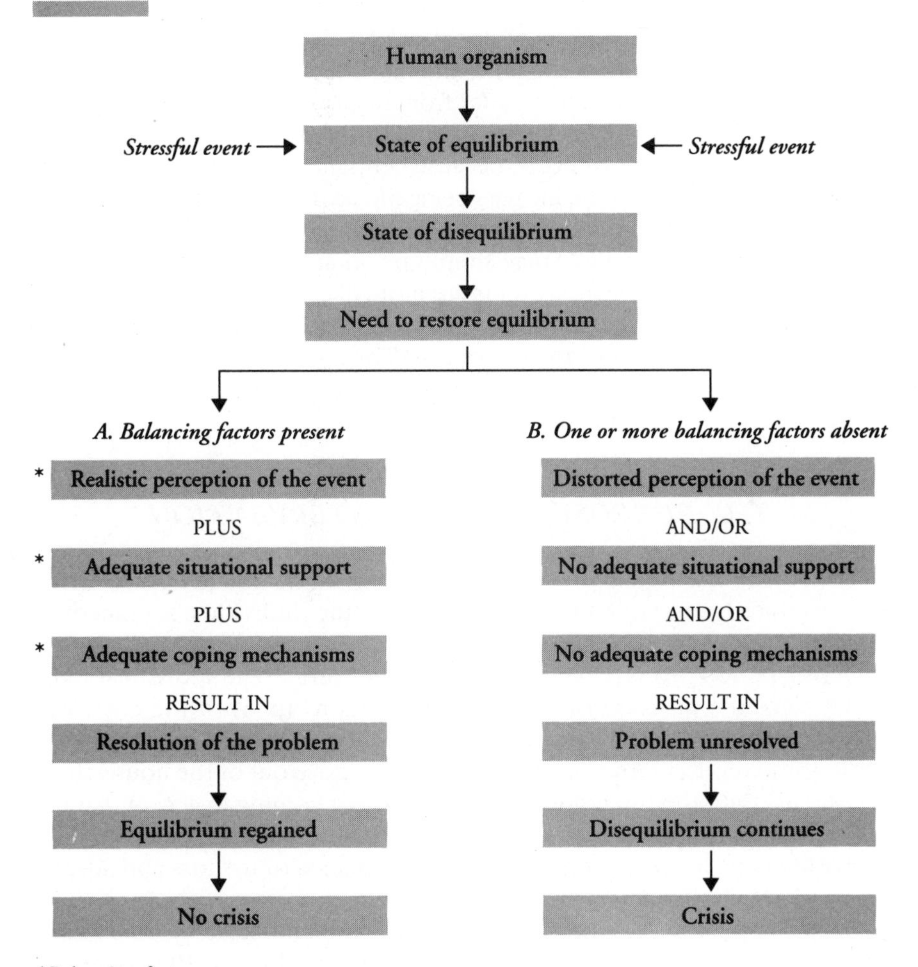

*Balancing factors

From D. C. Aguilera, *Crisis Intervention: Theory and Methodology,* 6th ed. Copyright 1990 by The C. V. Mosby Company. Reprinted by permission.

Aguilera (1990) has developed a paradigm (Figure 11.1) that shows what happens when a person in a state of equilibrium is confronted by a stressful event. Disequilibrium and the need to return to stability are felt. If the person perceives the event accurately and has adequate support and coping skills, the problem is resolved and crisis is avoided. In the absence of accurate perception, situational support, or coping skills, the problem remains unresolved and crisis results.

It is important to recognize that a given event, for example, losing one's job, may precipitate crisis in one individual but not in another. Whether losing a job will trigger a crisis depends upon a variety of factors, including the individual's skills for seeking a new job, his or her employability, his or her financial reserves, the reactions of significant others to the job loss, the other stressors in his or her life at the time, and his or her perceptions about each of these conditions (which may differ from reality). The centrality of career and the particular job to the person's sense of identity is yet another determinant of the potential for crisis, because centrality is related to the importance to the individual of the life goal being blocked. For a highly employable individual with good job-seeking skills, some financial reserves, and a supportive spouse, loss of a job may simply be an annoyance, not a crisis. In contrast, less serious negative events are more likely to precipitate crises for individuals with fewer coping skills, many stressors besides the precipitating event, and weak or absent support systems. The counselor cannot determine what may constitute crisis for a client by assessing whether the event would cause crisis in his or her own life.

THE PURPOSE OF CRISIS INTERVENTION

According to Aguilera (1990), "The minimum therapeutic goal of crisis intervention is psychological resolution of the individual's immediate crisis and restoration to at least the level of functioning that existed before the crisis period" (p. 21). The goal is a limited one: "The individual must either solve the problem or adapt to nonsolution" (p. 5). If the crisis was precipitated by loss of a job, then finding another job is a solution. If the crisis was precipitated by a spouse's decision to move out of the house, there is potential for either mending the relationship or coming to accept that the spouse is gone. In the case of the death of a significant other, the only alternative available for resolving the crisis is acceptance of the loss and adaptation to life without that person.

The focus of crisis intervention work is the immediate problem, not the totality of the client's personality or life issues. In fact, the counselor must guard against allowing the sessions to ramble onto other issues that might distract attention from resolving the event that has resulted in the client's disequilibrium. Crisis intervention is oriented in the present; it involves the client's developmental history only to illuminate an understanding of the specific crisis and to gain knowledge of coping abilities that have served the client in previous situations.

STRESSFUL EVENTS THAT MAY PRECIPITATE CRISES

The pioneering work in crisis intervention (Lindemann, 1944) was generated in response to the needs of survivors and family members of the

493 victims of a fire at the Cocoanut Grove nightclub in Boston. Lindemann's interventions anticipated the needs of these individuals, encouraged them to allow themselves to experience their losses, and provided support. Currently, the literature on crisis intervention includes both group and individual approaches for working with situational crises (e.g., loss of life or health) and developmental crises (e.g., the mid-life crisis). Some kinds of crisis (e.g., suicide) are common enough that they have been studied extensively and group programs have been designed that anticipate the crisis and focus on prevention. Our review of the literature identified at least thirty different kinds of occurrences that commonly overload human coping capacities and thus lead to crises.

Perhaps the best way of identifying the triggering events for crisis is to remember that crisis occurs when the client believes important life goals are blocked. And it is commonly stated that essentially all life goals relate to the love and work motives of human experience. Therefore, any event that jeopardizes success in either or both of these arenas has the potential to initiate crisis responses. Money and mortality, though not entirely independent of the love and work motives, are sometimes added as causes of crisis.

Table 11.1 provides a listing of many events that commonly precipitate crises in people's lives. It is important to realize that crisis is sometimes caused by an event that affects an important person in the client's environment. Any of the events listed in the table can seriously disrupt the quality and stability of life by depriving an individual of opportunities to love and be loved, to perform optimally in school or work settings, or to possess the necessities and amenities of a comfortable life. Though it could be established empirically that some of these events are regarded by people in general as much more serious than others, it is important to understand that one individual may be able to cope with the most egregious happening (e.g., the death of a child) with only temporary disruption of life activity. Even though the individual may experience a deep sense of loss, he or she is able to cope with the routines of life. Another person may be immobilized and propelled into crisis by an occurrence that might seem much less important to most people (e.g., failure on a school examination). Whether or not a crisis occurs depends not only on the severity of the loss, but also on the individual's coping resources.

Besides situational crises, crises may be precipitated by a range of developmental transitions, and these developmental crises are managed the same way. A developmental crisis occurs when an individual's coping skills fail under the stress of role transformations that are considered to be a usual part of maturing and aging. Examples include starting to school, moving to a new school, changing one's college major, entering the world of work, marriage, having children, retiring, and many more. Often these events are challenges that people seek and that can bring joy. However, if a client perceives himself or herself as unready for an event that he or she sees as timely or culturally expected, emotions of fear and guilt may immobilize him or her. The crisis that is experienced will closely resemble those created by the situations of loss presented in Table 11.1.

Table 11.1 Events That Trigger Crises

Type of Event	*Examples*	*Those Affected*
Death	Suicide Homicide Accidental death Natural death	Family Friends Associates Anticipated for self
Health and well-being	Physical illness Mental illness Injury Disability Abuse (physical, sexual, psychological) Substance abuse	Self, family, friends
Reproductive issues	Unwanted pregnancy Stillbirth Miscarriage	Self or significant others
Disruptions of intimate relationships	Arguments Infidelity Separation Divorce	Spouse, family, friends
Violence	Domestic Criminal (including rape) Civil	Self, family, friends
Disruptions of work or school	Layoffs Firings Strikes Academic failure	Self or significant others
Natural and environmental disasters	Storms Earthquakes Fires	Self or significant others
Financial emergencies and homelessness	Unexpected expense Bankruptcy Gambling losses Investment losses Theft or fraud	Self or significant others

STEPS IN CRISIS INTERVENTION

1. Establish a Helping Relationship

As Gilliland and James (1993) remind, "Basic listening and responding skills are the prerequisites for opening gateways into all other therapeutic modalities" (p. ix). In crisis intervention, it is crucial that the counselor develop a clear understanding of the event that precipitated the crisis and the meaning of that event to the client. At the same time, the client needs to feel the support that grows from being understood. Therefore, the basic relationship-building skills, that were discussed in Chapter 2, including active listening and the core conditions of empathy, positive regard, genuineness, and concreteness, form the bedrock on which crisis intervention is built. The fact that there is some urgency about restabilizing the client as quickly as possible does not reduce the necessity for employing these skills.

It will also facilitate a crisis interview if the counselor maintains calm confidence and hopeful expectation (Hersh, 1985). A calm and confident manner is reassuring to the client, who observes that the counselor is not overwhelmed by his or her problem. If the client is very emotional and out of control, it is sometimes necessary for the counselor to tell him or her to settle down and try to talk calmly so that the counselor can help figure out what to do. A direct statement such as "We will work something out that will help you face this situation better" bolsters hope. Hersh also recommends special attention to a comfortable environment, privacy, and the absence of time pressures.

2. Assure Safety

One of the first concerns about a client in crisis is how dangerous he or she may be to self or others (Aguilera, 1990; Gilliland & James, 1993; Hersh, 1985). The client may also be in danger from someone else. Among the conditions that may bring a client to a counseling office or motivate a telephone call for help are suicidal ideation or attempts, homicidal ideation, threatened or actual attacks upon self, and fear of attacking or hurting someone else. Fear that a loved one is in danger from a third party or from self may also motivate contact.

It is important to ask direct, specific questions about any of these circumstances. If the individual is planning to kill himself or herself, the counselor should ask when and how. Is the client merely thinking about it or is there a plan? How lethal is the plan? Does the plan include a time and place where the act is not likely to be discovered? Has the person been spending time alone brooding over problems? What is his or her support system like? There is no evidence that talking directly about suicide or homicide increases the likelihood that it will occur; in fact, talking about it may release some tension and reduce the likelihood. The decision to share the concern often represents an alternative to actually carrying out the act.

Of course, the counselor must judge the risk based on the answers to the questions and, if the situation is dangerous, take steps to involve family and other sources of support, to hospitalize the client, and to warn any intended victims. In situations where the safety of the client or another person is determined to be a concern, the counselor should seek consultation with a supervisor or colleague. Seeking a second opinion helps assure the best possible plan of action and alerts others in the agency or school that a crisis has occurred and that back-up assistance may be needed.

In instances of telephone contacts about family violence, Roberts (1991) also recommends that the counselor ask specific questions about whether the client is personally in danger, whether children are in danger, whether the attacker is present, whether police or medical personnel are needed, and whether the client wants to leave and can safely do so. A client will sometimes come to a counselor's office anticipating family violence or after an incident has occurred, and similar questions aimed at assessing the safety of all persons involved are necessary.

Children and adolescents are increasingly at risk from violent acts by peers, particularly attacks involving firearms. If an individual approaches a counselor with concerns about his or her own safety, the counselor should explore the situation carefully to assess the risk. In addition to working with the young person to defuse whatever conflict exists, it may be necessary to involve family and to notify appropriate authorities, especially if firearms have been seen in the hands of a potential attacker or other credible means of violence are identified.

Regardless of the nature of the destabilizing event, the client may be experiencing symptoms of stress that are sufficiently severe to disrupt normal patterns of eating, sleeping, and working. In such circumstances, consideration should be given to psychiatric evaluation for the purpose of prescribing appropriate medication for anxiety, depression, or psychotic symptoms.

3. *Conduct an Assessment*

In the assessment process, the crisis intervention worker secures information about the event that precipitated the crisis, what the event means to the client, the client's support system, and his or her functioning prior to the crisis (Aguilera, 1990; Hersh, 1985; Roberts, 1991). This information will help the counselor decide whether the consequences of the event might be moderated or reversed, whether the client's own coping skills can be mobilized to meet the challenge, who else might help and how, and what the counselor may need to do. Based on Caplan's (1964) germinal work on crisis intervention, Aguilera has developed a paradigm (Figure 11.1) showing how people react to stressful events that introduce disequilibrium.

The counselor should inquire first about what caused the crisis. Aguilera (1990) recommends opening with this simple direct question: "Why did you come for help today?" (p. 63). If the client skirts the issue by saying that he or she has been feeling upset for some time, it is important to persist in asking "Why today?" and "What happened that is different?" The purpose is to identify the "last straw" that overwhelmed the client's coping abilities; it might range in severity from a violent assault to rejection or humiliation (Roberts, 1991). The precipitating event usually has occurred within the last two weeks and often within the last twenty-four hours (Aguilera, 1990), but it may also "extend back as far as a few months or to an anniversary reaction of a major loss" (Hersh, 1985, p. 287).

The counselor working at crisis intervention employs concreteness (Chapter 3) and structuring and leading techniques (Chapter 6) to narrow the focus of the initial discussion to the precipitating event. Many clients have a myriad of chronic problems that preceded the crisis and may lead to follow-up counseling; for other clients, life may have been stable prior to the crisis. In either case, the purpose of crisis intervention is to restore the level of functioning that existed prior to the precipitating event. Involvement with other preexisting problems will only complicate and delay the planning of an intervention that will reduce the state of crisis.

While focusing on one problem, it is nevertheless important to encourage the client to expand on its personal impact. What feelings is the client experiencing (sadness, rage, panic, dread, embarrassment, guilt, etc.)? Has there been an impact on daily routines, sleep pattern, physical functioning, or relationships with others? How has the precipitating event threatened the client's life goals? Through discussion of these issues, the counselor seeks to understand the meaning of the destabilizing event to the client. "Unless the worker perceives the crisis situation as the client perceives it, all intervention strategies and procedures the worker might use may miss the mark and be of no value to the client" (Gilliland & James, 1993, p. 23). Though questioning is a necessary part of the assessment process in crisis intervention, the core listening skills of empathy, genuineness, and acceptance are also essential if the counselor is to gain access to the client's world of inner meaning. Remember also that feeling understood and accepted contributes directly to calming the client, instilling hope, and mobilizing his or her coping behaviors.

As the counselor comes to understand the meaning of an event to the client, it is necessary to "listen for and note cognitive distortions (overgeneralizations, catastrophizing), misconceptions, and irrational belief statements" (Roberts, 1991, p. 12). Premature direct confrontation of such distortions will lead to resistance and impede progress, but gentle attempts at cognitive restructuring may be tried. For example, a young man who has been dropped by his girlfriend may hold the views that he can't go on without her, is not a desirable person, and therefore is doomed to spend his unhappy life alone unless he gets her back. The counselor might respond, "Right now you are consumed with thinking about her and can do nothing else." The phrase "right now" very gently implies that this may not be a permanent condition. A more direct attempt at restructuring might be a question about whether he can name anyone other than the girlfriend who thinks he is nice. Although it would certainly be premature to attack or discount his girlfriend's action, the purpose of this question is to reduce any distortion that her decision reflects an accurate representation of his worth.

During the assessment process, the counselor observes the client's physical appearance, behavior, mood, speech pattern, attention span, and any signs of distress. The extent of the client's preoccupation with the crisis can be estimated from these cues.

Finally, it is important to develop an understanding of the client's functioning prior to the crisis. The purpose of this assessment is to determine how the client usually manages difficult situations and what skills have typically been available to him or her. Such a specific assessment of strengths provides direction as to what kinds of actions the client may be able to mobilize with the support and encouragement of the counselor. Brammer (1993, p. 21) lists the following as skills that the client may possess:

1. Perceptual skills (seeing problematic situations clearly, as challenging or dangerous, and as solvable)
2. Cognitive change skills (restructuring thoughts and altering self-defeating thinking)
3. Support networking skills (assessing, strengthening, and diversifying external sources of support)
4. Stress management and wellness skills (reducing tensions through environmental and self management)
5. Problem-solving skills (increasing problem solving competence through applying [decision-making] models to diverse problems)
6. Description and expression of feelings (accurate apprehension and articulation of anger, fear, guilt, love, depression, and joy)

The assessment process includes evaluating strengths in each of these six dimensions of coping skills. Plans for action should be designed to maximize the client's pre-crisis strengths and, where possible, to minimize dependence on skills that have not been a part of his or her repertoire. For example, one client may have a wide range of family and friends with whom to network for support, and another may be so isolated that emergency contact with a professional may constitute the bare beginning of a network. A given client may have perceived events and other persons accurately up to the point of the crisis and will usually be able to work out distortions related to the crisis with the counselor's help. A different client may routinely misinterpret the intent of others and perceive events in distorted ways; this client's habitual defensiveness will make it much more difficult to establish an accurate perspective on the event that precipitated the crisis.

4. Give Support

Assessing the client's support system involves finding out who in the client's environment cares what happens to him or her and has a favorable opinion of his or her worth. When self-esteem is low, calling upon such individuals to be attentive and provide comfort is important. An unmarried woman with an unwanted pregnancy will usually find her strength to cope with the crisis strengthened if the father and/or her own parents are able to assure her that she is still loved and that the crisis event has not deprived her of important relationships. Sometimes the contributions of persons in the support system may be more tangible, for example, providing financial support for needed medical attention. Gilliland and James (1993) are emphatic in their recommendation that the crisis worker directly express his or her caring for the client. Even if other support persons are scarce, the counselor has the opportunity to make it clear to the client that there is one person right here who really cares. When support persons are not evident in the client's daily life, plans should be made so that emergency contacts can be made with the counselor personally or with an emergency worker in off hours. This is, of course, especially important if there is suicidal ideation but insufficient evidence to consider hospitalization.

5. Assist with Action Plans

It is in the action-planning step that crisis intervention is probably most different from other forms of therapy. The client is in such a state of distress that some action step that will return him or her to a pre-crisis level of equilibrium must be identified. By definition, the client's own coping mechanisms have failed; therefore, the counselor must be willing to take an active role and will often be more directive than in other forms of counseling (Aguilera, 1990). Since the client's ego function has been inadequate to the task of defusing the problem, the counselor may be seen as temporarily "lending" his or her ego function, which is unimpaired by the experience of the crisis, to the client. By the time the action-planning stage of crisis intervention begins, the client is likely to have experienced some calming as a result of catharsis and of sharing the problem. The folk wisdom that "a problem shared is a problem halved" describes the impact of the preceding work. Because of the calming effect, the client's own coping abilities are likely to be more available to him or her than was the case at the beginning of the session.

Although the precipitating event and circumstances surrounding it are discussed as a part of the assessment process, it is necessary to help the client gain an accurate cognitive understanding of the crisis before seeking a solution (Aguilera, 1990; Gilliland & James, 1993; Hersh, 1985). The client's specific problem with the precipitating event—what consequence is so intolerable that the client cannot function—sets the parameters for determining what actions might provide relief. Since crisis intervention is not the time to try to resolve a myriad of concerns, the counselor must tenaciously hold the client's attention on one problem whose moderation will begin to restore equilibrium. It is often difficult to get the client to focus on one problem and to let go of the problems of some third party (e.g., a spouse) who is not present (Gilliland & James, 1993).

The search for possible actions begins with alternative ideas or solutions the client can think of. Even though many clients under stress may initially have a limited view of options, there are usually additional possibilities. Through the use of open-ended questions, the counselor tries to elicit, identify, and modify coping behaviors that have worked for the client before in similar situations. Some material that is initially at a minimal level of awareness for the client should become more available as the discussion of alternatives proceeds. When the client's ideas have surfaced, the counselor may add other possible actions to the list. Initially, it is useful to use brainstorming, where all possible actions that the two parties can think of are listed without evaluation. This process should expand the range of options and create the impression that there are many actions that have a possibility of making a difference.

Once alternatives have been listed, the counselor encourages the client to select one or more actions that he or she feels capable of accomplishing. The counselor "endeavors to develop a short-term plan that will help the client get through the immediate crisis, as well as making the transition to long-term coping" (Gilliland & James, 1993, p. 52). The counselor helps identify concrete positive actions that will help the client regain control of his or her life. The best plans are those that the client truly owns (Gilliland & James, 1993), but the counselor may have to give "specific directions . . . as to what should be tried as tentative solutions" (Aguilera, 1990, p. 64). Ideally, "the final part of the action plan involves cognitive mastery: restructuring, rebuilding, or replacing irrational beliefs and erroneous cognitions with rational beliefs and new cognitions" (Roberts, 1991, p. 12).

Among actions that may be appropriate are referrals to other sources of material assistance and support, such as housing, food, clothing, financial assistance, legal advice, or emergency contact. The counselor serves as a resource person to help the client in crisis find resources such as the Red

Cross, public assistance, legal aid, hotlines, and other community agencies. Counselors rarely manage these resources but should be networked with agencies that do so.

Before concluding a crisis session, it is important to judge whether the client's anxiety has decreased, whether the client can describe a plan of action on his or her own, and whether there is a glimmering of hope in the client's demeanor. It is also a good idea to readdress the questions of who else knows how the client has been feeling and whether the client is willing for the counselor to make direct contact with that person (e.g., spouse, parent, friend, or roommate). The help of another individual in expressing caring and providing support and a sense of hope can reduce tension for the client and encourage him or her to take the actions that have been planned.

6. *Arrange for Follow-up*

A follow-up meeting or telephone call should be arranged at a designated place and time to check on the client's progress toward resolution of the crisis (Roberts, 1991). Even though clients in crisis are usually well motivated to escape from the discomfort they are feeling, some plans are hard to execute and no plan comes with a guarantee of success. If the client has not begun to manage his or her problem by the time of the follow-up conversation, then recycling through any or all of the above steps may be in order.

Case: Carolyn

Carolyn, a 26-year-old single white female, sought counseling at a community mental health center. Transportation to the center was provided by a friend who stayed in the waiting room while the initial session took place.

In an initial flood of information, the client revealed that she "feels like she's in a box and can't get out," that she has never liked herself, that she feels like a "fat slob" even though her weight appeared near normal, and that she feels stupid even though she got good grades in high school and college. For the past six months, she had been forcing herself to vomit to try to gain control of her weight and has lost thirty pounds. Carolyn also said that she is tired all the time but even so has trouble sleeping.

When asked to explain what she meant by "feeling like she is in a box," she explained that she has very rigid views of what is right and wrong and that she constantly finds herself doing things that are against her beliefs. She was especially disapproving of her active sex life with a variety of partners.

She has maintained employment as a medical records clerk and reported that her employer is satisfied with her performance, though she feels that she always has loose ends hanging around and should be more efficient. The client's language and behavior in describing her circumstances had a dramatic quality, punctuated by both nervous laughter and tears. She reported feeling very depressed and that recently she has been feeling like taking her life sometimes.

Initially, the counselor responded using mainly reflective responses that captured the meaning and feeling of the client's statements and open-ended questions to elicit more information. After much of the presenting situation had been disclosed, the counselor asked the client what brought her to counseling *at this time.* The client reported that she had been feeling like she was "increasingly out of control" and that she became frightened as she began dwelling on a suicide plan.

The counselor inquired about the plan and learned that the client had been stockpiling tranquilizers and had read that when these are taken with alcohol they can lead to sleep from which one does not awake. Furthermore, she lives alone, has no contact with her family, and has only the one friend who has brought her to the center. The likelihood that she would be discovered before death was not high. The counselor asked about the men with whom she had intimate relationships and learned that these were sexual encounters more than relationships and that it would be pure chance if she heard from one of them during a suicide attempt. When asked whether she would consider relinquishing her supply of tranquilizers to her friend for safekeeping and to be dispensed in small numbers, the client declined. The counselor concluded that the client was at considerable risk of suicide.

The counselor also reasoned that the client's vomiting, combined with the control issues that are typical of eating disorders, constituted a second source of risk. The counselor decided that the risk of allowing the client to return home was one that should not be taken. Indeed, the client's own statement that she was frightened about what she might do indicated that some tangible action was needed.

Clearly, two medical problems existed. First, an evaluation was needed to determine what medication would help with the client's depression. Tranquilizers are not a usual treatment for depression and constitute a source of danger with this client. Second, the potential of an eating disorder must be evaluated. The counselor carefully explained that each of these problems were medical and that evaluation at the psychiatric unit of a local hospital would be the safest course of action. Though Carolyn felt anxious about going to the hospital, she agreed to go if her friend would accompany her. The counselor and client discussed what the friend should be told about the referral, and it was agreed that she should be told that the client was in danger and needed an immediate evaluation. The friend agreed to take the client directly to the hospital, and a telephone referral to the hospital was made.

The psychiatrist recommended in-patient treatment to more fully evaluate the client's mood, to fully explore her eating pattern, and to establish the client on antidepressant medication. The client declined to be hospitalized on a Friday afternoon but became frightened for her safety by Sunday and admitted herself to the hospital, where she was retained for treatment for two weeks. She was subsequently referred to an eating disorder specialist and continued to see the hospital psychiatrist on an out-patient basis to deal with her depression.

Questions for Further Thought

1. Identify ways in which this crisis intervention counseling is similar to and different from the generic model of counseling.
2. Do you agree with the decision that the counselor made to seek immediate referral? Why or why not?
3. If the client had not chosen to cooperate with the referral, what actions would be open to the counselor?
4. If the indications of suicidal intention had been lower, what changes would you recommend as to actions to be taken?

SUMMARY

This chapter has addressed how a counselor should proceed to help a client who presents with a problem so distressing that it has destabilized the client's equilibrium and rendered him or her incapable of acting to resolve the problem. Many of the skills that are evident in all counseling are evident in crisis intervention work: the core conditions of effective listening and responding are employed, the client's environmental support system is reviewed, the client's strengths are identified, and problem analysis and action planning take place. However, little effort is devoted to developing a history of the client, and resolution of pre-crisis emotional concerns is not among the crisis intervention goals. The process begins with careful listening that allows for problem identification and catharsis as well as sharing of the burden. It is useful to identify early in the session exactly what the event was that caused the client to lose control of his or her coping abilities. Care must be taken to ensure the physical safety of the client and any others who may be in danger from the client. Together, client and counselor search for alternative plans of action, based on the client's coping skills in previous situations similar to the one that precipitated the crisis. The counselor is often more active in suggesting alternatives and structuring the discussion than he or she would be in other counseling. Some plan of action must be agreed upon within the session. Finally, it is necessary to follow up with the client to make sure that action has been taken and that it is beginning to moderate the crisis and to restore pre-crisis abilities to deal with the challenges of living.

REFERENCES

Aguilera, D. C. (1990). *Crisis intervention* (6th ed.). St. Louis: C. V. Mosby.

Brammer, L. M. (1993). *The helping relationship* (5th ed.). Boston: Allyn and Bacon.

Caplan, G. (1964). *Principles of preventive psychiatry.* New York: Basic Books.

Gilliland, B. E., & James, R. K. (1993). *Crisis intervention strategies* (2nd ed.). Pacific Grove, CA: Brooks/Cole.

Hersh, J. B. (1985). Interviewing college students in crisis. *Journal of counseling and development, 63,* 286–289.

Lindemann, E. (1944). Symptomatology and management of acute grief. *American Journal of Psychiatry, 101,* 141–148.

Roberts, A. R. (1991). *Contemporary perspectives on crisis intervention and prevention.* Englewood Cliffs, NJ: Prentice-Hall.

Part Three

Theoretical and Professional Issues

CHAPTER 12

MAJOR THEORIES OF COUNSELING

The early chapters of this book presented a generic model of counseling. That model traced the flow of the counseling process as it typically occurs, from relationship building and initial disclosure through deeper exploration to action planning. We emphasized elements of counseling that are accepted by most professional counselors. Because we have blended compatible elements from several theories, some would call our model eclectic.

This chapter examines some of the *differences* in the counseling process that arise from the different theoretical positions on counseling. Each theory includes certain assumptions about the nature of human beings, and these assumptions influence the counseling interventions that adherents of a particular theory will choose to promote client progress. Because of their varying emphases, each theory elaborates more on certain aspects of the generic counseling model. The special contributions of each theory are discussed in the material that follows.

If you are using this book as the sole source in a first counseling course or volunteer training setting, the brief summaries of the several theories will also serve to elaborate on the theoretical foundations of the book presented in Chapter 1 and to suggest a path for further reading. If you are using the book in combination with a theories of counseling text, this chapter will assist you in linking your more extensive study of the theories to the generic model.

The chapter begins by presenting a structure that shows a relationship among the theories on a continuum according to, among other criteria, the emphasis on affect or cognition in the counseling process. The first section also provides some historical perspective on the development of counseling during the twentieth century. Introductions to six of the more important counseling theories follow. We acknowledge that the task of condensing the significant elements of six theories of counseling into a single chapter is challenging, and thus the presentations are necessarily rather skeletal. A serious student of counseling will want to study a comparative theories text

(Burke, 1989; Corey, 1991; Corsini & Wedding, 1989; Lynn & Garske, 1985; Patterson, 1986), as well as original works by the key authors of each of the theoretical approaches.

STRUCTURE FOR RELATING COUNSELING THEORIES

The counseling profession traces its origin to the work of Frank Parsons (1909), who was concerned about helping young people make effective career choices in a work environment that had become very diversified by the Industrial Revolution. He regarded career decision making as primarily a rational process of guided self-appraisal, analysis of work opportunities, and matching self with opportunities. From 1909 until the 1940s, much of the progress made by the counseling profession involved the development of better means of assessing people's aptitudes and interests through testing and the collection and publication of occupational information. In many respects, counseling became more "rational" and more scientific during that period. E. G. Williamson (1939) and others described the decision-making process as an application of the scientific method. The counselor, as an expert in measuring skills and abilities as well as in sharing career information, served the function of directing clients toward the careers for which they were best suited. This rational approach to counseling came to be known as *directive counseling* because of the counselor's role in advising clients about career choices. It was also referred to as *trait-factor counseling* because personal traits were matched with factors needed for success in various jobs. Though not totally ignored, clients' feelings about themselves were treated as second in importance to their thoughts, and "irrational" feelings were to be controlled rather than worked through.

In 1942, a book by Carl Rogers entitled *Counseling and Psychotherapy* was published; it was destined to change the counseling profession profoundly. Rogers described an approach to counseling that assumed that the client had capabilities within himself or herself for knowing best how to manage his or her affairs. The role of the counselor was to facilitate the client's process of self-exploration and personal growth by providing a nurturing relationship within which the client could express self freely and develop new insights. This approach came to be called *nondirective counseling,* because the counselor was not an adviser. The client was seen as self-directing, and the counselor had no advice to give. The client's feelings about self were seen to be the central concern of counseling, because these feelings were the elements of self-concept. According to Rogers, one's concept of self strongly influences how one responds to the challenges of life.

For twenty years after the publication of Rogers' initial work, a battle raged within the counseling profession over which of the viewpoints, directive or nondirective, was "right." Slowly many counselors came to the view

Figure 12.1 *Continuum of Counseling Theories*

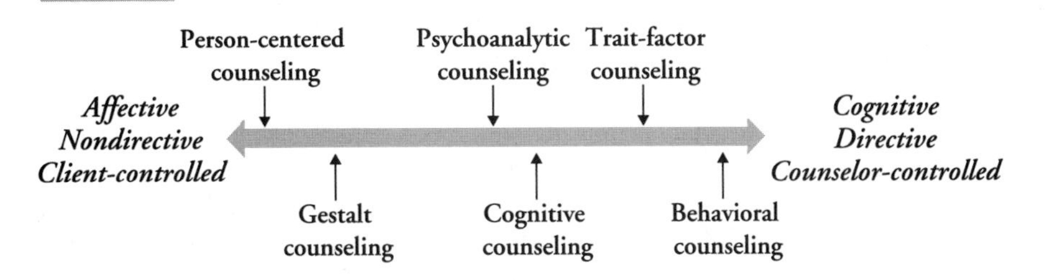

that there were elements of truth in each viewpoint, that some clients might be helped most by the one approach and others by the alternative, and that many aspiring counselors were more adept at one approach or the other. Furthermore, since Rogers' approach had opened the door to the inclusion of the full range of human experience as appropriate material for counseling, counselors began looking toward theories that had previously been regarded as within the territory of psychotherapists. Counseling and psychotherapy began their journey toward merging into a single discipline. Ironically, even though Rogers was as vocal in his objections to Freudian therapy as to Williamsonian counseling, the work of Sigmund Freud and his psychoanalytically oriented descendants became incorporated into the counseling literature largely as a result of Rogers' expansion of the horizons of the problems with which a counselor could help. *Freudian counseling* (also referred to as psychoanalytic or dynamic counseling) took up a position between the rational and affective viewpoints; its practitioners regarded both thoughts and feelings as important. Other theories that have been subsequently developed lie on a continuum whose extremes are defined by the initial arguments over directive and nondirective counseling.

C. H. Patterson (1986), after reviewing a number of classifications systems for counseling theories, concluded that contemporary approaches could be arranged on a continuum with primary emphasis on cognitive process at one extreme and primary emphasis on affective concerns at the other. Figure 12.1 shows a continuum like that proposed by Patterson, with the theories to be discussed in the remainder of this chapter placed in appropriate locations. Generally, those counselors who focus heavily on affect are also nondirective and see the client as controlling the content of the sessions. Counselors who are more cognitive in orientation are more inclined toward directive approaches, in which the counselor is more in control of content of the session. We have chosen to present summaries of six theoretical positions that are most frequently cited as germinal thinking on which counseling practice is based and from which other theoretical variations arise. Although their placement on the continuum cannot be regarded as precise, it was purposeful.

Person-centered counseling (originally called client-centered) places great emphasis on affect and client autonomy and thus appears at the extreme left. *Gestalt counseling* was placed to the right of person-centered counseling because the counselor is somewhat more inclined to manipulate conditions in order to elicit exploration by the client. *Psychoanalytic counseling* is placed near the center of the continuum because it emphasizes both affect and thought processes. The right side of the continuum shows those theories that emphasize cognitive process, with each system being progressively more directive and counselor-centered as one moves toward the right end of the continuum. *Behavioral counseling* appears at the extreme right because it pays little attention to affect and is heavily counselor-controlled.

PERSON-CENTERED COUNSELING

Carl Rogers (1942, 1951, 1961, 1980, 1986) is known as the founder of the person-centered approach to counseling. Two other names, nondirective counseling and client-centered counseling, were attached to this approach in Rogers' earlier writings, and a majority of the references to the system are to be found under "client-centered counseling" or "client-centered therapy." The change to a person-centered approach reflects Rogers' (1980) later recognition that his system worked in any setting in which a helper sets out to promote human psychological growth and that many of those who are helped (e.g., students in a classroom) do not think of themselves as clients.

Rogers' approach to the helping process was presented initially as an alternative to psychoanalytic psychotherapy, in which he was first trained. Because his views of human nature (1942) appealed to educators and his method of counseling did not require extensive psychological training, the person-centered approach was adopted by many then practicing counselors and had a great influence on the preparation of new counselors. Rogers' work is regarded as one of the principal forces in shaping current counseling and psychotherapy.

The Nature of Humans

In person-centered counseling, human beings are seen as possessing positive goodness and the desire to become "fully functioning," that is, to live as effectively as possible. This view of human nature contrasts with the Freudian view that people possess impulses that, if not adequately socialized, will lead to behavior that is destructive to self and others. According to Rogers, if a person is permitted to develop freely, he or she will flourish and become a positive, achieving individual. Because of the faith in human nature expressed in Rogers' theory, it is considered a humanistic approach to counseling.

Person-centered counseling is based on a theory of personality referred to as *self theory.* One's view of self within the context of environment influences one's actions and personal satisfactions. If provided with a nurturing environment, people will grow with confidence toward self-actualization—becoming all they can be. If they do not receive the love and support of significant others, they will likely come to see themselves as lacking in worth and to see others as untrustworthy. Behavior will become defensive (self-protective), and growth toward self-actualization will be hampered.

An important principle of self theory is the belief that a person's perceptions of self and environment (including significant others) *are* reality for that person. Thus, if an individual sees self as incompetent or parents as mean, he or she will act on that belief, even if others view him or her as brilliant or the parents as kind. Telling an underachieving student that he or she is capable seldom makes much difference, because the assessment probably conflicts with the student's personal reality. Personal reality may be changed through counseling, but usually not by such a direct intervention as substituting the judgment of the helper for that of the client.

The Counseling Process

Because Rogers viewed humans as positive and self-actualizing by nature, he conceived the counselor's role as providing conditions that would permit self-discovery and that would encourage the client's natural tendency toward personal growth. The nurturing conditions described by Rogers are essentially those presented in Chapter 3 as vital to the initial-disclosure stage of counseling. If the counselor is totally accepting of the client as a person, relates empathically to the client's reality, and behaves in a genuine way (behavior is congruent with feelings), the client will be free to discover and express the positive core of his or her being. As the client comes to perceive self more positively in this nurturing environment, he or she will be enabled to function more effectively. The counselor not only provides a nurturing environment that may be missing elsewhere in the client's life, but also functions as a role model of how a fully functioning person relates with others.

The underlying philosophical view of human nature is far more important to the practice of person-centered counseling than is any particular set of techniques or any body of knowledge. It has been said that in person-centered counseling, the helper learns how to *be* a counselor rather than how to *do* counseling. The counselor must be comfortable enough within himself or herself to become fully involved with the world of the client without fear of losing his or her own sense of wholeness. And the counselor must care enough about his or her clients to be willing to experience their pain. Through all this, the counselor must be able to retain his or her own sense of separateness and emotional perspective on the client's difficulties. Since the client is seen as having the potential to solve his or her own problems, the counselor is not perceived as having expert knowledge to share with the client.

The person-centered counselor functions with the techniques that are toward the least leading end of the continuum of lead (see Table 6.1, p. 106). There is extensive use of silence, acceptance, restatement, and clarification, with the client taking the lead on what is discussed and being responsible for outcomes. If a need for information is perceived in the course of a discussion, the counselor encourages the client to seek information outside the counseling session. A purely person-centered counselor would not be likely to use tests, although a counselor who uses some person-centered procedures might include some testing at the client's request. Person-centered counselors encourage careful self-exploration, but they tend to avoid confrontation and interpretation as tools for hastening insight. There is little focus on specific action planning except as initiated by the client. It is assumed that as the client becomes free to actualize his or her potential through the exploration process, behavior change will occur naturally and without prompting from the counselor.

Contribution to the Generic Model of Counseling

The work of Carl Rogers has made a great contribution to the generic model of counseling that is presented in this book. The obvious contribution is his clear description of the helping relationship that forms the substance of the first stage of the counseling process. Because of Rogers' work, counselors have learned to become better listeners. Even counselors who prefer more active and counselor-initiated methods have come to recognize the importance of employing relationship-building conditions to encourage the clients to reveal significant elements of their personal reality. It has been said that counseling must begin where the client is—and to learn where the client is, most counselors employ methods that Rogers defined.

A second important contribution, which owes its origin to Rogers' statements about the nature of humans, is the view that clients are ultimately responsible for their own lives. Even though some counselors may not be as optimistic as Rogers in believing that all persons are fundamentally good, most counselors recognize that the counselor cannot and should not attempt to control a client's actions (except in instances where physical safety is a real concern). Counselors of essentially all theoretical persuasions now maintain that they are working to help clients achieve their own goals.

Person-centered counseling, then, is based on a positive view of human nature that is consonant with American views of freedom and self-determination. Rogers has described a methodology that can be learned relatively easily by persons who are themselves "fully functioning." Because Rogers and his colleagues have demonstrated through a strong research program that person-centered counseling works with a wide variety of clients, many of his ideas have been incorporated into other counseling models. For example, they are an important element of the generic model of counseling presented in this book.

GESTALT COUNSELING

Gestalt counseling was developed by Frederick (Fritz) Perls, who, like Carl Rogers, was educated as a Freudian psychotherapist. Though a dynamic and effective counselor, Perls was not as diligent as Rogers in writing about his work, and there are probably more conceptual "loose ends" in his system. When considered together, several publications by Perls and others (Fagan & Shepherd, 1970; Perls, 1969; Perls, Hefferline, & Goodman, 1951; Polster & Polster, 1973; Van de Riet & Korb, 1980; Yontef & Simkin, 1989) provide an adequate basis for understanding his contributions.

The Nature of Humans

The name of this approach to counseling comes from the German word *Gestalt,* which can be roughly translated as "whole." A central concept of this view of human nature is that the whole is greater than the sum of the parts and that the organism functions as a whole. In addition, humans are seen as constantly striving for balance in their lives, this balance being threatened by events outside of self as well as by internal conflict. Regardless of the source or nature of threats to the individual's balance, the whole organism is disrupted when imbalance occurs. Thus, difficulty in one's marriage will influence one's job performance; failure in school is likely to affect interpersonal relationships; and success in sports will impact on friendships. The concept of the whole also applies to interaction between physiological and psychological aspects of a person; thus, emotion often gets expressed through physical means. Understanding nonverbal behavior is important to understanding the whole person.

Perls believed that instinct plays a part in motivating behavior, but he emphasized a "hunger" instinct rather than sexual instinct as in the Freudian theory. This hunger instinct motivates persons to "take in" elements of the environment. Although this instinct drives the organism, it is not seen as bad or good. In fact, Gestalt theorists tend to see behavior not as "bad" or "good" but as "effective" or "ineffective."

As persons live and experience, they often develop elements of personality that differ and that "want" different things. For example, one's "fighting self" might want to battle with perceived enemies, but one's "loving self" wants to make peace. An abused child might seek revenge against the abuser at one time and seek only to be loved at another time. According to Perls, an important part of understanding any human being is understanding the polarities that exist simultaneously within the personality. Part of the helping process is encouraging the client to resolve these polarities and to work toward a more unified set of motives.

Gestalt counseling is placed toward the affective and client-centered end of the continuum of counseling theories, because client feelings are emphasized and because the client is seen as responsible for his or her own coping behavior. As in person-centered counseling, the counselor is seen primarily as a facilitator of the client's growth toward self-responsibility; the private reality of the individual is regarded as the basis for his or her interactions with the world at large; and the growth of awareness, which alters that private reality, results in more effective functioning. Gestalt counseling, like person-centered counseling, is classified as a humanistic approach, but different emphases among its adherents result in different counseling procedures, as will become clear in the next subsection.

The Counseling Process

In the Gestalt system, the purpose of counseling is to encourage personal growth. Neurosis is seen as an interruption of the growth process, and defensiveness as a process that pulls energy away from living effectively. Growth is seen as a sequence of occurrences that proceeds from experience (with a life situation) to sensing (taking in the qualities of the situation through the senses) to excitement (becoming involved with the event) to Gestalt formation (integrating the experience into one's stored perspectives of the world). Stagnation occurs when a person uses defenses to isolate self from new experiences.

Some Gestalt counselors work to frustrate the client's attempts to hide from new experience during the counseling session by pointing out when the client retreats into defensive behavior. Others are less directly confrontative, but nevertheless encourage their clients to consider discrepancies that occur in their thoughts and feelings about life and that lead to the blockage of experience and growth. Either way, confrontation is an important counseling technique, though most contemporary Gestalt counselors are decidedly more gentle in their approach than Perls was.

Frequently, confrontation is based on discrepancies between the client's verbal statements and his or her nonverbal behavior. For example, it is common for people to laugh when embarrassed in an attempt to dismiss the embarrassment. A Gestalt counselor would observe, "You say you were embarrassed but you are laughing." Confrontation is also used in response to discrepant components of verbal statements the client makes and to omissions—what the client is not saying (e.g., about the significant others in his or her life). Through this kind of intervention, the client is moved closer and closer to an authentic awareness and acceptance of his or her own experiences.

A number of techniques used by Gestalt counselors focus on bringing "then and there" experience into the "here and now." If a client is concerned about a disagreement with his or her spouse, the client may be asked to reenact some aspect of the dispute, alternately taking the roles of self and spouse. Instead of *talking about* what has occurred, the client reexperiences the occurrence, and the affect is recreated. New understandings often lead to new behaviors or new acceptance of self or others. When such reenactments of life events reach back to earlier phases of the client's life, the process is referred to as "finishing unfinished business." For example, a young adult may bring the residue of conflict he or she had with parents during the growing years into relationships with legitimate authorities of adult life, such as employers. By reprocessing and "digesting" the old conflict, such a client is freed to interact more objectively with persons currently present in his or her life.

Gestalt counselors use many techniques, or *experiments,* to bring the client into clearer touch with himself or herself. Only a few of these will be described here. Role playing is perhaps the most important. This can be done with the counselor playing the part of a significant other and the client playing himself or herself, with the counselor playing the client and the client playing the significant other, or with the client playing both roles. Sometimes the client is asked to change chairs as he or she shifts roles. Another variation of the "empty chair" technique occurs when the client is asked to play separate parts of self, perhaps the "assertive" self and the "weak" self. With all role-playing techniques, the counselor permits only enough talking about the situation to be able to structure roles and then moves to bring the action into the "here and now."

Still other experiments are the exaggeration game and the "I take responsibility" game. In the first of these, the client is asked to exaggerate a view that has been expressed and that the counselor sees as defensive. As the affect and content become exaggerated, the client comes to see the inaccuracy and may take back part of the original defensive statement. The "I take responsibility" game works in much the same way. The client is instructed to repeat a questionable statement and follow it with the words "and I take responsibility for what I have said." If the client in retrospect has doubts about the original statement, he or she must alter it in order to feel able to take responsibility for it.

Though Gestalt counseling is not interpretive—that is, the counselor does not interpret behavior—the Gestalt counselor needs diagnostic skills in order to recognize clients' defensive strategies. In order to make observations that cause a client to look into self, the counselor must be able to recognize the client's defensive attempts to hide. As the counselor works with a given client, themes emerge that outline the client's characteristic ways of coping with life, and particularly with other people. The counselor calls the client's attention to attempts to hide within counseling sessions and may refer to such defensiveness as "phoniness." The counselor continually encourages the client to move toward authentic behavior that reflects his or her internal state accurately, while recognizing that such authenticity can be very difficult.

As counseling proceeds and the client finishes unfinished business and approaches an understanding of his or her authentic self, he or she is able to establish effective contact with other people and participate more fully in daily life experiences. The growth process is contagious, in the sense that success in life situations builds the client's belief that he or she can safely reach out rather than hide. At some point, the client's tendency to reach out exceeds his or her tendency to hide, and the growth process reaches a momentum that makes further counseling unnecessary.

Contribution to the Generic Model of Counseling

Gestalt counseling is a client-oriented approach that suits the temperament of counselors who wish to intervene more actively than is typical in the person-centered approach. Much of the leading is characteristic of the second stage of counseling, as described in Chapter 4, but the Gestalt counselor stops short of interpretation (which is more characteristic of systems further toward the active end of the continuum of theories). One of the important contributions of Gestalt counseling to the generic model of this book is its elaboration of many techniques that can help the client see self more clearly. The client is still in charge of the outcome, but the counselor can use very active techniques in promoting self-insight.

Because of its attention to the client as a total being, the Gestalt approach also extends the counselor's focus to the physiological response to experience. The Gestalt counselor will frequently ask a client to be aware of the tension or relaxation in his or her muscles, the responses of the circulatory or digestive system, or his or her overt and observable physical behavior as he or she discusses stressful (or happy) life events. Training in Gestalt work will enhance any counselor's ability to integrate cues from the client's nonverbal behavior into the counseling process, whether or not the counselor chooses to confront such behavior directly.

Finally, we reiterate that Gestalt counseling is a growth-oriented approach that, like the Rogerian approach, emphasizes human potential. The counselor attempts to remove barriers in order to permit the client the freedom to become more fully functioning. Unlike person-centered counseling, Gestalt counseling includes more statements about the nature of barriers and is thus more diagnostic in its orientation.

PSYCHOANALYTIC COUNSELING

Psychoanalysis was originated by Sigmund Freud, who developed his theory from his experience as a therapist and wrote about his work for a period of nearly fifty years prior to 1939. Strachey (1953–74) edited a multivolume edition of Freud's complete works, facilitating reference to this important germinal material. Psychoanalytic theory is also sometimes referred to as *dynamic theory* because of its emphasis on the interactions (dynamics) of unconscious processes ". . . where conflicts arise between the need for tension reduction and the inhibition of basic instinctual drives" (Baker, 1985, p. 25).

As we indicated earlier in this chapter, Freud's works were not initially considered important to the practice of counseling. His work focused on the alleviation of serious emotional problems; counseling in its early years centered on decision making, mainly about career choices. The introduction of person-centered counseling opened the door for counselors to deal with a broader range of human concerns and emotions, and the counseling profession has subsequently moved steadily toward increased involvement with clients with emotional disturbances. Counselors have embraced psychoanalytic theory as an important stream of thought, and many psychoanalytic concepts (e.g., the unconscious, defenses, ego) have, in fact, entered the common language.

Freud's form of psychoanalysis was a very thorough, long-term helping process that placed heavy emphasis on the client's historical psychosexual development; the goal was for the client to gain insight on all aspects of his or her personality. Today, few practitioners, even psychoanalysts, practice that kind of treatment. Modern psychotherapists generally question whether such total analysis of the personality is necessary for a majority of clients, and the time and financial investment in such a process exceeds most people's resources. Consequently, shorter-term counseling based on psychoanalytic theory is now a much more common form of treatment, and many counselors who practice with one or more other theories as a base also use some psychoanalytic principles.

It would be difficult to overestimate Freud's contributions to the understanding of the human psyche and to the process of helping people resolve emotional problems through talking about them. Burke (1989) points out that much of the theoretical basis for today's psychotherapy is either a further development of Freud's work (Adler, 1927; Alexander, 1963; Erikson, 1963; Fromm, 1941, 1976; Jung, 1954; Sullivan, 1953) or a reaction against it in the form of the person-centered, Gestalt, cognitive, and behavioral systems. Psychoanalytic counseling and psychotherapy account for a very large proportion of helping services.

The Nature of Humans

Freud saw humans as biological beings driven by an instinctual desire for personal pleasure (gratification). The life force, or *libido,* was postulated as the energy source that propels people toward behavior that satisfies the pleasure motive. Only through the process of socialization are humans redirected toward behavior that allows satisfaction of personal needs in ways that are not destructive or unacceptable to others. If allowed to grow and develop without controls, people would serve their own selfish pleasures without regard for the rights of others or the accomplishment of useful work. There is no element in Freudian theory related to any tendency on the part of humans toward self-actualization: rather, humans are seen as operating by the pleasure principle and in need of shaping toward positive endeavors.

Psychosexual Development. In Freud's theory, pleasure is linked with sexuality, and libido is a driving force toward gratification. Freud postulated that the desire for sexual pleasure is a lifelong drive that begins in infancy and is first satisfied by sucking the mother's breasts. This first stage is called the oral stage; if the baby's oral needs are not met, later greediness or acquisitiveness may result (Corey, 1991). Anal, phallic, latency, and genital stages each follow in turn, bringing needs that, if satisfied, allow growth toward psychological maturity. If the child's needs are not accepted by the parents and are not satisfied in acceptable ways, the probable consequence will be "fixation" on meeting these needs at a later period in ways that are not effective.

Freud's views on sexuality have resulted in a great deal of controversy over the years. Many people object to the idea that all pleasure has a sexual component and that sexuality is experienced from birth. To many, the idea that humans are sexually motivated beings who begin seeking satisfaction at birth is offensive. Some have seen Freud as not only incorrect but perverted. Others see truth in the theory of infantile sexuality but place less emphasis on sexual pleasure as the primary motivator of all human behavior. Many women have seen his views of female sexuality, heavily influenced by the social strictures of the nineteenth century, to be offensive.

Whether or not one chooses to accept Freud's ideas about psychosexual development, there is much to be learned from his work on the unconscious, the structure of personality, and the defense mechanisms (summarized next). Unfortunately, many counselors have discounted this important work because his views on sexuality seem extreme to them or because his views on women and mental health, developed in Victorian times, do not pass muster today. We suggest that instead of rejecting Freud's contributions, it is wise to use his work selectively, in those areas where it can be applied in a contemporary context.

The Unconscious. One of the most important ideas introduced by Freud was the concept that people are unaware of much of their mental processes—that mental activity can be unconscious. A person's unconscious motivation is based on instinct as modified and socialized by interaction with significant others, mainly parents, during the formative years. If satisfaction of instinctual needs is blocked through ineffective parenting so that acceptable means of expression cannot be found, then unconscious motivators will propel the individual to satisfy those needs by whatever means available. The key concept is that people frequently do not understand why they behave as they do, since motivation can be unconscious. This may be true even when a person seems to have a plausible explanation for particular behaviors. The plausible explanation may simply be a socially acceptable defense covering a motive of which the person is not aware. All behavior is understood to be purposeful (as a means of satisfying drives) though the individual may not be aware of the purposes. Discovering motives and developing effective means of meeting needs is one of the tasks of counseling.

The Structure of Personality. Freud postulated that the personality is made up of three interacting components: the id, the ego, and the superego. The *id* is the source of psychic energy and the locus of instinct. It propels the individual toward the search for pleasure, without regard for consequences. The *ego,* often referred to as the "executive" of the personality, is in touch with external reality and through experience gains strength to help the individual satisfy needs in acceptable ways. The ego is the seat of rational thought. The *superego* consists of learned principles of right and wrong and serves also to control instinctual gratification. When born, the individual is motivated by an unfettered id, instinctively seeking immediate gratification of needs. As time passes, the ego develops a repertoire of strategies for coping with life so as to satisfy needs in positive ways, and the superego develops a store of attitudes about right and wrong that regulates behavior. Both the ego and the superego establish control over the direct expression of id impulses.

Contemporary practitioners of psychoanalytic counseling tend to emphasize the development of the ego as the key to effective functioning. This leads to a sociocultural perspective on development, which emphasizes the role of interpersonal relationships and cultural institutions in the development of normal and pathological behavior (Burke, 1989).

Defense Mechanisms. When an individual is confronted with demands for which the ego has no solutions, anxiety results. The person becomes afraid because, without ego-mediated solutions, he or she may directly express unacceptable impulses. The resulting behavior may be both ineffective and embarrassing. In such circumstances, ego defenses come into play to soften the blow on the ego and reduce stress. For example, a student who is failing in school may use the defense of denial to dismiss the problem or the defense of rationalization to explain why he or she can't do better. A defense mechanism works in the sense that it takes the pressure off the ego. Unfortunately, if defense mechanisms are used repeatedly, the result is that the person dismisses many demands to perform and thus misses many opportunities to succeed at life tasks. Neurosis is said to occur when an individual uses defense mechanisms in interaction with other people to such an extent that few or no rewarding relationships are experienced. Psychosis results if the ego becomes so overwhelmed that its contact with external reality is severed and distorted thought patterns result.

Defense mechanisms are employed as a part of the unconscious process of a person's mental functioning. Therefore, it would rarely be useful to tell a client "You are just being defensive." It is important for counselors to be able to recognize defenses and to help clients explore troublesome circumstances to find coping responses that can replace the defensive ones. It is also important to realize that there are times when everyone needs the temporary respite that a defense provides and that moderate defensiveness in the face of tough circumstances is healthy and necessary.

Common defense mechanisms include denial, rationalization, intellectualization, projection, and regression. It is beyond the scope of this chapter to define and describe these and other defense mechanisms; you are urged to refer to virtually any general psychology text for further information.

Freud described humans as instinctually motivated beings, seeking pleasures that have sexual roots. Development through the psychosexual stages in a nurturing environment allows the individual to develop a healthy personality in which the ego is strong. When the ego is not strong enough to face life's challenges, defenses are developed to protect the ego. Defenses serve a useful purpose, provided they are not overused as substitutes for coping with challenges. The dynamic process of personality includes unconscious thought; thus, people are not fully aware of all the motives behind their behavior.

The Counseling Process

Fundamental to the psychoanalytic counseling process is the belief that humans relegate material they cannot tolerate to the unconscious, using defense mechanisms (e.g., repression). Since crucial issues have been pushed out of awareness without being resolved, unmet needs keep intruding into the fabric of life. The process of counseling, then, encourages the client to dislodge unconscious material and resolve the conflicts contained therein.

The client is encouraged to talk as freely as possible about troublesome situations. Talking about these issues often leads to the recall of related thoughts that were repressed. In some cases, free association is used. In *free association,* the client is asked to suspend control over what he or she says and just let speech flow—regardless of how disconnected or bizarre the material seems. Sometimes dreams are also analyzed for clues to the unconscious.

Regardless of the method of disclosure—problem discussion, free association, or dream analysis—the counselor tries to understand the client's motives and to interpret the client's behavior for him or her. The counselor depends on his or her knowledge of psychodynamics to lead the client toward new insights. The counselor also uses events in his or her own relationship with the client as samples of the client's behavior that can be interpreted (see the description of transference in Chapter 10). There is often substantial discharge of emotion by the client (referred to as catharsis), as painful circumstances are explored and new insights are achieved.

With interpretation serving as an important counselor lead, psychoanalytic counseling depends heavily on the counselor's knowledge of personality dynamics. Psychoanalytic counseling places the most emphasis on the second stage of counseling—in-depth exploration.

Contribution to the Generic Model of Counseling

The greatest contribution of the psychoanalytic approach to the practice of contemporary counseling is the theory of personality and its application to the diagnostic process. The structure of personality as posited by Freud provides a convenient framework for analyzing human functioning. It helps the counselor consider the comparative contribution of impulse, reason, and conscience to the behavior of a given client. Along with the tripartite structure of personality comes the concept that the ego develops strength through positive experience with the external world (environment). At times, the strength of the ego is not equal to the demands placed on it, and the client becomes defensive. The nature of defense mechanisms, the purposes they serve, and the problems they cause are all issues that are important to the diagnostic process. Transference, countertransference, and resistance are manifestations of the process of ego defense (see Chapter 10).

Elements of Freud's original methods of conducting therapy certainly survive today. Many counselors use interpretation as a predominant lead, attempting to help clients see their experiences using a psychodynamic structure of personality. Such counselors work heavily with the client's reports of his or her experiences in everyday life and also use observations of the client's interaction with the counselor as material for interpretation. Dream analysis and free association are used less as material for the interpretive process than they once were, but they remain a window on the unconscious experience of the client. Much of psychoanalytic counseling is in-depth exploration, with comparatively less emphasis on relationship building or action plans. However, all three stages of the counseling process will receive some attention from most psychoanalytic counselors.

We have placed psychoanalytic counseling near the middle of the continuum of counseling theories because it focuses on both cognitive and affective material and both counselor and client are active in the process of counseling. A psychoanalytic counselor will not be content to have a client talk about feeling without meaning, or meaning without feeling. It is the client's responsibility to reveal self as freely as possible. It is the counselor's responsibility to interpret the client's experiences so that the client will gain greater insight and the ego can increase its coping capacity, directing effective rather than defensive responses to environmental demands.

COGNITIVE COUNSELING

Cognitive counselors regard erroneous thinking as the source of emotional upset and ineffective behavior. Events occur in everyone's life that involve loss, disappointment, and failure to accomplish valued goals. Cognitive therapists believe that people who are able to think effectively about their experiences are able to put negative events in perspective and get on with life, and those who do not think effectively tend to perseverate on negative happenings and allow them to disrupt their happiness and effectiveness.

Albert Ellis (Ellis, 1962, 1973, 1977; Ellis & Bernard, 1986), founder of rational-emotive psychotherapy, is probably the best known of the cognitive therapists. Like Rogers and Perls, he was trained as a psychoanalytic therapist, but he came to believe that the traditional approach was inefficient and that the process sidetracked clients from learning how to live more effectively. Influenced by learning theory, Ellis began to develop a new approach to counseling in which clients are taught to think rationally about blocks to accomplishing love and work goals. Psychiatrist Aaron Beck (Beck, 1972, 1976; Beck, Rush, Shaw, & Emery, 1979) and psychologist Donald Meichenbaum (1977, 1985) have both gained recognition for their development of related theories of cognitive counseling and psychotherapy.

The Nature of Humans

For cognitive theorists, humans are thinking beings with the capacity to be rational or irrational, erroneous or realistic, in their thinking. According to Patterson (1986), "Cognitive therapy is based on the commonsensical idea that what people think and say about themselves—their attitudes, ideas, and ideals—is relevant and important" (p. 33). All cognitive therapists subscribe to the view that what people *think* about the experiences they have determines how they *feel* about those experiences and what they will *do.* Ellis, Beck, and Meichenbaum all posit internal dialogue that mediates a person's reactions to stressful events.

Ellis explains that negative emotion and ineffective behavior are the results of irrational thinking. It is not the events in people's lives that create bad feelings but how they think about those events. For example, suppose a person is snubbed at a party by someone he or she thinks is attractive. Such an event might be unpleasant for almost anyone, but it becomes a problem, according to *rational-emotive theory (RET),* when the snubbed individual "catastrophizes" about the event. The individual may have such "irrational" thoughts as "I can't stand being snubbed" or "It is an absolute necessity that I be loved and approved by others almost all the time." If the person simply thinks "It's too bad that person snubbed me; I'd like to spend time with him or her," then negative emotion will not get out of hand and the person can plan to work toward another opportunity for contact.

However, if the person catastrophizes about the incident, negative emotion and ineffective behavior will result. Energy will be expended in feeling sorry for self and possibly either avoiding contact with or planning retaliation against the other person. Clearly, such behavior will not achieve the desired result of having further opportunity to spend time with the person.

The RET view of personality is often referred to as an ABC theory, in which A is an activating event, B is the person's thought about the event, and C is the emotional and behavioral reaction. If the thinking at B is irrational, the emotional reaction will be negative and the behavior is likely to be inappropriate and ineffective for accomplishing the desired outcome.

There are many similarities between Beck's and Ellis's views of how humans come to behave ineffectively. Ellis has listed a dozen specific irrational thoughts any one or more of which may lie at the source of an individual's difficulties. Examples include "[I] must be unfailingly competent and perfect in all [I] undertake," and "It is horrible, terrible, or catastrophic when things do not go the way [I] want them to go" (Ellis, 1977, p. 10). Ellis has identified all the statements as applying to love and/or work motives, and in that sense he repeats Freud's view of human priorities. It is easy to see how extreme ideas like those just quoted would result in feelings that one could not be successful and behaviors that are not well designed to bring about success. Beck (1972) has focused more on the nature of erroneous thought processes than on specific life events. He identified four patterns of erroneous thinking:

- Dichotomous thinking—believing that everything is either good or bad, black or white, with nothing in between
- Overgeneralization—arriving at far-reaching conclusions on the basis of little data
- Magnification—overestimating the importance of an event (essentially the same as the RET concept of catastrophizing)
- Arbitrary inference—drawing conclusions that things are bad with no evidence

Ellis speaks of "rational and irrational thoughts," Beck of "automatic thoughts," and Meichenbaum of "self-instructions"—all of which are spontaneous thought processes that occur when an individual is confronted with experience. These derive from adults' instructions that children internalize while growing up and apply to new situations. Thus, it may be seen that parenting practices influence subsequent ability to think effectively, to feel confident and competent, and to behave using the maximum amount of one's resources. For clients to change their behavior, they must learn new ways of thinking, which is the means by which cognitive counseling achieves its purpose.

The Counseling Process

The counseling process has similar elements in each of the cognitive systems, and cognitive restructuring is the principal mechanism of change. The first step is to have the client describe the stressful situations in his or her life and to identify the faulty thinking that underlies the feelings. The counselor identifies the irrational thoughts, automatic negative thoughts, and silent assumptions that the client uses to interpret (erroneously) his or her experience. The cognitive errors and distortions may be explained to the client, as frequently occurs in RET, or discussion is structured so that the client comes to see his or her errors of reasoning. Then more adaptive alternative patterns of thinking can be developed. In RET, the alternatives are relatively easy to construct once the irrational thinking is identified because they are direct modifications of Ellis's list of irrational thoughts—which reduce the compelling nature of the client's thinking to a more reasoned version. For example, the thought "I must be loved and accepted by almost everyone" might be modified to "It would be nice to be widely cared for, but there are some people who matter who like me and I can get along okay even if I don't get the attention I want from some others." Beck's cognitive approach leads to examination of the client's story for examples of dichotomous thinking, overgeneralization, magnification, and arbitrary inference in the client's response to troublesome circumstances. Seeing that a client looks at behavior in an all-or-nothing manner, makes "mountains of molehills," or jumps to conclusions, the counselor tries to help the client understand the erroneous nature of his or her thinking. Meichenbaum also suggests that it is important to look for self-talk that, if present, would lead to better conclusions. Thus, it is important to consider what the client has overlooked as well as the cognitive errors that he or she has made.

It was stated above that RET is an ABC theory of personality, where A is the precipitating event, B is the client's thought about the event, and C is the emotional and behavioral response. The counseling process adds a DEF component (Ellis & Bernard, 1986), where D is the "disputing intervention." The scientific method is used to identify new rational thoughts about the client's situation. When this occurs, the client arrives at E, a new "effective philosophy"—which in turn makes a new set of "feelings," F, possible. Ellis has referred to this process as "depropagandizing" because it results in the client's giving up irrational beliefs that have been taught to him or her (as propaganda) during the formative years. Techniques such as persuasion, suggestion, instruction, and discovery of new ways of thinking through the Socratic method are common in the cognitive therapies. Planning of specific actions to take place outside of counseling, rehearsing the client's role in these new actions, and reviewing success are also parts of the cognitive counseling process. Thus, clients are given homework assignments to experiment with their environment between sessions and acquire new learnings.

The cognitive approach to counseling is placed near the rational and counselor-controlled end of the continuum of counseling theories. The counselor enters the client's world of experience through his or her thinking (cognitive) processes, and the counselor takes charge of the counseling. Nevertheless, it is the client's goal of coping more effectively with troublesome experience that shapes the content for counseling. Although Ellis places little importance on first-stage counseling skills, the other cognitive therapists see the conditions of the first stage as important to establishing a therapeutic alliance with the client, creating a climate of trust so that the client will respond to the interventions suggested by the counselor. None of the cognitive therapists would see the first-stage conditions alone as sufficient for effective and efficient treatment.

In cognitive counseling, the in-depth exploration process of the second stage results in the identification of issues with which the client is experiencing difficulty and an examination of the thought patterns that underlie the unpleasant feelings and ineffective behaviors. The counselor helps the client identify fallacies in perceptions, inaccuracies in information, and self-defeating behaviors. Although the affect attached to certain circumstances signals where the client is experiencing difficulty and the severity of that difficulty, discussion is focused more on thoughts and actions than on feelings. The client's goal of becoming more effective in managing troublesome aspects of his or her life becomes clearer. Some cognitive counselors (e.g., Ellis) tend to move fairly quickly through this stage; others engage in more discussion, and their exploration process may not seem very different from that of a person-centered or psychoanalytic counselor, except for the emphasis on thoughts.

The third stage is more elaborate in the cognitive approach than in others. The client is instructed to go out and behave differently, either by implementing newly discovered rational thinking or possibly by experimenting with finding new information about his or her beliefs about others. Scientific problem solving, led by the counselor but with the client as an active participant, leads to plans of action. As with any new learning, new patterns may not be implemented perfectly at first, and reinforcement and refinement are necessary. Cognitive counseling aims to bring about changes in actions in a comparatively short time.

Contribution to the Generic Model of Counseling

Cognitive counseling provides an additional model for understanding and intervening in human behavior. The point of entry is through the thinking process and the assumption that more effective thinking will result in more satisfactory (to the client) feelings and behavior. For certain clients, the identification of faulty thinking and the teaching of more effective ways of viewing life experiences can result in rapid improvement. The RET approach in particular bypasses a complicated historical diagnostic process and moves directly to supporting change in the client. Cognitive restructuring gives the client direct help in changing self-defeating thoughts and feelings. Just as person-centered counseling provides the clearest view of the first stage of counseling, cognitive counseling—with its emphasis on action planning—provides the strongest descriptions of the third stage. In addition, it adds a diagnostic process based more on learning theory than on personality theory.

TRAIT-FACTOR COUNSELING

As described in the introduction to this chapter, trait-factor counseling was considered to be *the* counseling method for many years during the early part of the twentieth century. Although originated by Frank Parsons (1909), the trait-factor approach was more clearly articulated by E. G. Williamson (1939, 1950, 1965). Because Williamson was a professor at the University of Minnesota, the method is sometimes referred to as the Minnesota school of counseling. Although a number of newer humanistic, psychodynamic, cognitive, and behavioral approaches to counseling have emerged to help people with emotional and behavioral problems, the trait-factor approach remains the most common method for assisting people with educational and vocational choices.

The Nature of Humans

Proponents of the trait-factor approach to counseling do not take a position that human beings are either innately good or innately bad. They share the view of their cognitive colleagues that what people learn serves as the basis for their behavior. People are treated as rational decision makers, and children are seen as naive and inexperienced—in need of adult guidance toward effective decision making. Each individual is endowed with a unique set of abilities and develops particular skills, interests, and attitudes that have implications for decision making. The better an individual knows his or her particular characteristics and the demands of the world, the more effective he or she will be in planning and having a good and useful life.

The Counseling Process

Trait-factor counseling is sometimes referred to as counselor-centered counseling, because so much of the process depends on the counselor's activity. The counselor uses his or her expert skills to help the client *objectively* assess various traits that have implications for problem solving and decision making. This objective assessment includes the use of test scores that reflect the client's skills, abilities, and interests. The counselor tells the client how he or she compares with other persons on specific scores so that he or she can choose a course of action that will lead to success and satisfaction. The counselor actively introduces information about the world of work and about educational opportunities so that the client can become more aware of alternatives. Guidance about such things as how to study and how to get along with others is also offered.

Williamson described the counseling process as a six-step sequence including analysis, synthesis, diagnosis, prognosis, counseling, and follow-up. Each of these steps refers to the counselor's role. Analysis is collecting all relevant data about the client, including test scores and information from school records if available. Synthesis is the process of bringing all such data together into a comprehensive picture of the client. Diagnosis, based on the synthesis of the information collected, indicates where any impediment to progress or decision making may lie—dependency, lack of information, inaccurate self-concept, choice anxiety, or no problem at all. Prognosis consists of predicting further developments relating to the client's progress. Counseling is helping the client take steps that will bring about adjustment or readjustment. These steps include conforming with societal expectations, changing environments, selecting new environments, learning new skills, and changing attitudes. Finally, follow-up is the process whereby the counselor watches the client's progress and repeats the other steps if the client is not developing positively. From this description, it is clear that the system is directed toward helping individuals find their spot in the larger scheme of things. It is also implied that, at first, the counselor will often see the best course of action more clearly than will the client.

Problem solving in the trait-factor approach also follows the scientific method: a problem is clearly defined, data (information) is collected, courses of action are defined, one course of action is selected and implemented, and evaluation takes place that may lead to the repeat of the cycle. The appeal of this approach lies in the belief that people can figure out how to handle their lives better. Generally, if emotional disturbance is not a significant factor in the client's problem, the process of carefully reviewing a problem or decision with the help of a counselor results in clarification of the choice to be made and improvement in decision-making skills. Trait-factor counseling was devised for work with developmental and decision-making problems, and Williamson advised that more disturbed clients should be referred to other helping professionals, such as psychologists. In contemporary counseling settings, most counselors have training in trait-factor methods as well as one or more of the other approaches described in this chapter, and they will choose specific techniques depending on the nature of the client's presenting concern.

Contribution to the Generic Model of Counseling

Trait-factor counseling is the method of choice when the problem faced can be resolved by rational decision making and when the level of emotional complication is low. This is the case for many clients who seek help with educational and vocational choices. The trait-factor approach provides the counselor with tools for collecting objective data about the client and about opportunities that may be available to the client. This information, often stored in and sorted by computer, is fed into the client's decision-making process.

There is little emphasis on the first or second stages of the generic model of counseling presented in this book. Relationship building is seen as rapport building and serves the purpose of making the client more receptive to the counselor's information and advice. In-depth exploration is concentrated on a systematic exploration of client traits in order to inform the problem-solving and decision-making process. Goals that are directly related to the problem or problems the client presents are discussed and clarified. The counseling interventions are primarily those of the third stage, which emphasize action steps. The counselor is active in formulating some of the action steps and may offer advice if he or she considers the client to be lacking the perspective to make good choices on his or her own.

We note that a basic philosophical position stated by Williamson is in conflict with the generic model presented in this book. He believed that counselors are responsible for perpetuating societal values. Such a viewpoint supposes that all counselors know what society values and that what society values is good for everyone. Since humankind is in a constant state of finding new truths to replace old ones, since different cultures hold different values, and since there are many different ways of living a happy and productive life, we find it inappropriate for counselors to espouse a particular value system in their counseling. The role of helping clients find values that work in mediating conflicts between personal freedoms and the rights of others is a much more complex activity than Williamson construed it to be.

However, the importance of information (knowledge) in making good decisions is widely accepted and is probably the major contribution of trait-factor theory to contemporary counseling. Modern computer and psychometric technology has greatly increased the availability and accuracy of information that may be used for decision making. In counseling settings that focus on career decisions and job search strategies, trait-factor counseling enjoys a renewed emphasis, and there is the capacity for greater thoroughness of analysis than was possible before. In most settings where counselors work, trait-factor counseling will be useful for some problems of some clients.

BEHAVIORAL COUNSELING

Behavioral counseling is based on learning theory. The fundamental assumption is that all behavior is learned and therefore can be changed by implementing strategies to produce new learning. The personality is regarded as the product of accumulated learnings.

The purpose of behavioral counseling is to change ineffective and self-defeating behavior, and only measurable behavior change is regarded as evidence of successful counseling. Generally, behavioral counselors do not regard hypothetical concepts about mental functioning, such as the unconscious, as important to the counseling process. Self-understanding is not an outcome goal.

No single author is credited with the development of behavioral counseling. Joseph Wolpe's (1958) work on reciprocal inhibition applied the principles of classical conditioning to changing neurotic behavior. B. F. Skinner (1971) is widely recognized for his work in developing operant conditioning techniques used in behavioral counseling, although he was not a therapist himself. Together with modeling, operant conditioning and classical conditioning are the principal methods employed in behavioral counseling. Lazarus (1989) and Kazdin (1980) offer more contemporary applications of behavioral methods. Interest in behavioral methods increased during the late 1960s, when many people became disenchanted with Rogerian methods as a predominant approach to counseling. More recently, narrowly conceived behavioral approaches have declined in popularity, and some behavioral counselors (e.g., Meichenbaum, 1977) have turned their attention to the thought processes that mediate behavior, blending their work into that of cognitive counselors. Lazarus (1989) has described a broad array of behavioral techniques, each of which provides clients with new opportunities for learning strategies of self-management.

The Nature of Humans

Behaviorists see human behavior as a function of heredity and environment. This view is often called deterministic, because both elements that shape behavior are largely beyond the control of the individual. One is born with certain inherited equipment that cannot be changed, and so the only variable left to alter after birth is the environment. What one learns from the environment determines behavior; changing the environment changes behavior.

Behaviorists hold no general view that humankind tends toward good or evil. Given adequate hereditary characteristics, any individual can become good or evil depending upon what he or she learns from the environment.

Constructs such as the self-concept, the ego, and the unconscious have no meaning in describing human nature in a strict behavioral system. Behaviorists do not necessarily deny that such mechanisms exist, but say that, if they exist, it is impossible for the counselor to observe or manipulate them. The description of humans as capable of learning is sufficient to behavioral counseling. A lot of knowledge about how people learn exists, and it makes sense to use it to influence them toward effective behavior.

The Counseling Process

Goal Setting. Behavioral counseling places great emphasis on clear definition of goals. Goals are stated in terms of behavior change so that observation will provide evidence that can be measured. A goal such as "I'd like to get along better with my parents" would not be acceptable. A more specific goal, such as "I will be home for dinner at least four nights a week to share some part of my life in pleasant conversation," would be seen as a step toward a better relationship with parents. Because the goal is specific behavior, counselor and client are able to assess the extent of accomplishment. Krumboltz (1966) stated that many nonbehavioral counseling efforts fail because of a lack of specific enough goals.

Frequently, clients are referred for counseling by significant others who are dissatisfied with their behavior. Behavioral counselors are perhaps more amenable than other counselors to the suggestions of significant others about the need for clients to change. For example, a child may be referred to counseling because she fails to meet the curfew standards of her parents. A behavioral counselor might center his or her work on changing the unwanted behavior without devoting a lot of time to understanding the client's affective experiences in historical relationships with parents or peers.

The client is usually provided with the opportunity to participate in the goal-setting process, even when problem behaviors are quite obvious to the counselor from the outset. In some instances, the client may come with his or her own goals in mind, as in the case of a person seeking to gain control over eating habits.

Since goals are specific, the counselor and client have direct means for documenting change. It is possible to identify and count specific *target behaviors* that are to be eliminated or increased as a result of counseling. The frequency of occurrence of the target behavior at the outset of counseling is considered to be the *baseline* against which progress is measured.

Strategies for Change. Counseling strategies are based on the principles of learning. The client is taught to think differently about his or her behavior (in cognitive-behavioral approaches) or simply conditioned to behave differently. Operant conditioning is one of the most common procedures used in behavioral counseling. The procedure, which can be used to eliminate undesirable behaviors or to develop positive behaviors, uses reinforcement techniques. If the counselor is attempting to help the client eliminate an undesirable behavior, he or she first determines what environmental conditions are supporting the behavior and then arranges for the reinforcers to be eliminated. A child who acts out at home or in school is frequently seeking the attention of parents or teachers. Often parents and teachers pay attention to the child only when he or she is misbehaving. In an operant conditioning plan, the counselor would teach parents or teachers to withhold attention to misbehavior and to provide attention instead when the child does something positive, such as doing his or her chores or homework. If significant others consistently reward positive behavior with attention and fail to respond to negative behavior that is attention seeking, the client will learn new behaviors that succeed in attracting the attention he or she wants. In operant conditioning, the client's behavior is selectively reinforced to increase desired behaviors in a variety of ways—for example, through positive attention, free time after completion of tasks, candies, etc. Undesirable behaviors may be discouraged through negative consequences, such as isolation or withholding of privileges.

Desensitization training, used to help clients reduce or eliminate irrational fears or phobias, is based on the principles of classical conditioning. First, the client is asked to be as specific as possible about the condition that produces the anxiety, such as being in high places. A list is developed that arranges frightening conditions on a hierarchy from least frightening (e.g., standing on a chair) to most frightening (e.g., standing at the edge of a cliff though safely behind a railing). The client is then taught to relax his or her body through breath control and muscle control. When completely relaxed, the client is asked to think about the frightening circumstances, starting with the least frightening, while the counselor continues to encourage relaxation. Eventually, the client can tolerate the more frightening thoughts while still remaining relaxed. The feared circumstance becomes paired with relaxation and the positive feelings that accompany relaxation. Finally, the client can be encouraged to experiment with real feared circumstances while practicing self-relaxation techniques.

Modeling is yet another process whereby the client is taught new behaviors. A model performs in a situation with which the client has difficulty, and the client observes his or her behavior. The model may also be asked to think aloud as he or she is performing, giving the client access to the thought processes that lead to behavioral consequences. This procedure may be applied more informally by placing the client with effective models in real-life situations, such as work or school. Public television's Fred Rogers is very effective in using modeling in his TV program for children.

As we have mentioned previously, cognitive restructuring is closely related to behavioral methods in that it attempts to change behavior by changing how one thinks. However, it is more appropriately considered above with cognitive counseling.

The sampling of behavioral procedures presented here is far from exhaustive. The common thread in these and other behavioral strategies is the establishing of conditions for new learning to take place. This often requires the manipulation of conditions in the client's life external to counseling, and significant others may become involved in consultation on how to support behavior change in the client.

Contribution to the Generic Model of Counseling

Behavioral counseling places very little emphasis on the history of how a problem has developed, except for an assessment of the learning conditions in the environment that have sustained an unwanted behavior or failed to support a desired one. This approach depends on learning theory, rather than personality theory, as a basis for understanding a client's behavior. In that respect, it is substantially different from the other aproaches discussed in this chapter, with the exception of its close relative, cognitive counseling. Self-understanding and insight into developmental issues are not a focus of behavioral counseling.

The process of behavioral counseling moves quickly from the first stage (initial disclosure) to the third stage (action planning). The first stage is accomplished without special emphasis on empathy, acceptance, or genuineness. These conditions are simply helpful in learning what the client's problems are. Once the problems are identified, goals are set rather quickly and as specifically as possible. Goals all address specific target behaviors, and behavioral counselors emphasis the development of specific goals more than any other system of counseling. The major emphasis of behavioral counseling lies in third-stage strategies that develop commitment to action, often through environmental manipulation. The procedures of behavioral counseling are very effective in relieving irrational fears (phobias), in modifying the behavior of children who have difficulty adapting to and making use of classroom environments, and in helping clients overcome various addictions. Counselors whose primary allegiance is to approaches that are more insight-oriented often use behavioral methods for these specific problems.

We have presented behavioral counseling in its most conservative conception in this chapter to show how much it can contrast with some of the other theoretical orientations presented earlier. However, it is important to understand that behavioral counselors are as concerned and committed to helping their clients as are other counselors and that they are aware of the affective states of their clients. A common perception that they are cold and mechanistic toward their clients is unfounded. Their procedures are dictated by their belief that the most effective means of helping is through setting conditions for new learning rather than through extensive discussion of feelings rooted in developmental experiences.

SUMMARY

A variety of theoretical emphases in counseling relate to the generic model of counseling presented in Chapters 2 through 5 of this book. Each places emphasis on conditions or techniques appropriate to one or more of the stages of the generic model, while deemphasizing conditions or techniques of the other stages. Person-centered counseling emphasizes first-stage conditions almost to the exclusion of third-stage techniques, and behavioral counseling emphasizes third-stage techniques almost to the exclusion of first-stage conditions.

Counseling theories can be arranged on a continuum from client-controlled and affective in orientation to counselor-controlled and cognitive in orientation. In presenting our generic model of counseling, we acknowledge that each of the more doctrinaire systems makes an important contribution in detailing procedures that are important to one or more of the stages. We believe that a student of counseling will benefit by understanding the generic model first and then adding techniques and procedures that can be learned by studying the various contributing theories. Through such a process, the counselor-in-training first gains a broad view of available theory and technique and then chooses and practices what fits best with his or her personal style. Many counselors will retain an eclecticism based on the generic model; others will gravitate to one system as the predominant source of diagnostic and helping tools. Most counselors recognize that certain kinds of client concerns respond better to one of the different counseling approaches.

REFERENCES

Adler, A. (1927). *The practice and theory of individual psychotherapy.* New York: Harcourt.

Alexander, F. M. (1963). *Fundamentals of psychoanalysis.* New York: Norton.

Baker, E. L. (1985). Psychoanalysis and psychoanalytic psychotherapy. In J. L. Lynn & J. P. Garske (Eds.), *Contemporary psychotherapies* (pp. 19–67). Columbus, OH: Charles E. Merrill.

Beck, A. (1972). *Depression: Causes and treatment.* Philadelphia: University of Pennsylvania Press.

Beck, A. (1976). *Cognitive therapy and emotional disorders.* New York: International Universities Press.

Beck, A., Rush, A., Shaw, B., & Emery, G. (1979). *Cognitive therapy of depression.* New York: Guilford.

Burke, J. F. (1989). *Contemporary approaches to psychotherapy and counseling: The self-regulation and maturity model.* Pacific Grove, CA: Brooks/Cole.
Corey, G. R. (1991). *Theory and practice of counseling and psychotherapy* (4th ed.). Pacific Grove, CA: Brooks/Cole.
Corsini, R., & Wedding, D. (1989). *Current psychotherapies* (4th ed.). Itasca, IL: Peacock.
Ellis, A. (1962). *Reason and emotion in psychotherapy.* New York: Lyle Stuart.
Ellis, A. (1973). *Humanistic psychotherapy.* New York: Julian Press.
Ellis, A. (1977). The basic clinical theory of rational-emotive therapy. In A. Ellis & R. Grieger (Eds.), *Handbook of rational-emotive therapy: Vol. 1.* New York: Springer.
Ellis, A., & Bernard, M. E. (1986). What is rational-emotive therapy (RET)? In A. Ellis & R. Grieger (Eds.), *Handbook of rational-emotive therapy: Vol. 2* (pp. 3–30). New York: Springer.
Erikson, E. (1963). *Childhood and society* (2nd ed). New York: Norton.
Fagan, J., & Shepherd, I. (1970). *Gestalt therapy now.* New York: Harper Colophon.
Fromm, E. (1941). *Escape from freedom.* New York: Holt, Rinehart and Winston.
Fromm, E. (1976). *To have or to be.* New York: Harper & Row.
Jung, C. G. (1954). *Collected works: The practice of psychotherapy* (Vol. 16). New York: Pantheon.
Kazdin, A. E. (1980). *Behavior modification in applied settings* (rev. ed.). Homewood, IL: Dorsey.
Krumboltz, J. D. (Ed.). (1966). *Revolution in counseling.* Boston: Houghton Mifflin.
Lazarus, A. A. (1989). *The practice of multimodal therapy.* Baltimore: Johns Hopkins University Press.
Lynn, J. L., & Garske, J. P. (1985). *Contemporary psychotherapies.* Columbus, OH: Charles E. Merrill.
Meichenbaum, D. (1977). *Cognitive behavior modification: An integrative approach.* New York: Plenum.
Meichenbaum, D. (1985). *Stress inoculation training.* New York: Pergamon.
Parsons, F. (1909). *Choosing a vocation.* Boston: Houghton Mifflin.
Patterson, C. H. (1986). *Theories of counseling and psychotherapy* (4th ed.). New York: Harper & Row.
Perls, F. (1969). *Gestalt therapy verbatim.* Moab, UT: Real People Press.
Perls, F., Hefferline, R., & Goodman, P. (1951). *Gestalt therapy: Excitement and growth in human personality.* New York: Dell.
Polster, E., & Polster, M. (1973). *Gestalt therapy integrated.* New York: Brunner/Mazel.
Rogers, C. R. (1942). *Counseling and psychotherapy.* Boston: Houghton Mifflin.
Rogers, C. R. (1951). *Client-centered therapy.* Boston: Houghton Mifflin.
Rogers, C. R. (1961). *On becoming a person.* Boston: Houghton Mifflin.
Rogers, C. R. (1980). *A way of being.* Boston: Houghton Mifflin.
Rogers, C. R. (1986). Carl Rogers on the development of the person-centered approach. *Person-Centered Review, 1,* 257–259.
Skinner, B. F. (1971). *Beyond freedom and dignity.* New York: Knopf.
Strachey, J. (Ed.). (1953–74). *The standard edition of the complete psychological works of Sigmund Freud.* London: Hogarth.
Sullivan, H. S. (1953). *The interpersonal theory of psychiatry.* New York: Norton.
Van de Riet, V., & Korb, M. (1980). *Gestalt therapy: An introduction.* New York: Pergamon.
Williamson, E. G. (1939). *How to counsel students.* New York: McGraw-Hill.
Williamson, E. G. (1950). *Counseling adolescents.* New York: McGraw-Hill.

Williamson, E. G. (1965). *Vocational counseling.* New York: McGraw-Hill.
Wolpe, J. (1958). *Psychotherapy by reciprocal inhibition.* Stanford, CA: Stanford University Press.
Yontef, G. M., & Simkin, J. S. (1989). Gestalt therapy. In R. J. Corsini & D. Wedding (Eds.), *Current psychotherapies* (4th ed.) (pp. 323–361). Itasca, IL: Peacock.

CHAPTER 13

ETHICS IN COUNSELING

As earlier chapters have demonstrated, effective counseling requires that a client be capable of self-disclosure and self-exploration and motivated to change attitudes and behaviors. Successful counseling also demands a skilled, empathic, and trustworthy counselor to guide and support the client through the change process. Counselors who violate their clients' trust, are insensitive to clients' needs and values, fail to see clients as individuals, or experiment with counseling interventions for which they have no training or experience are acting unethically. In each of these situations, the counselor is not making the best interests of the client the highest priority and is instead serving some other purpose—usually self-interest. Whenever this happens, the counselor's behavior is defined as unethical. For example, suppose a counselor called Linda shares confidential information from a counseling session with friends at a party because "it makes such a funny story." Linda is putting her own needs to impress her friends ahead of the welfare of the client. One partygoer may recognize the client's name and decide not to offer the client a job, or another may repeat the story so that eventually the client hears it from some stranger and feels betrayed by the counselor he trusted. This client may even resolve never to visit another helping professional again, regardless of his distress. In several ways, then, negative effects may come to a client when the counselor's self-interest is placed above the client's welfare.

Counselors have two other broad ethical obligations: to be loyal to the institution that employs them, and to promote the good reputation of the counseling profession. Linda, the talkative counselor, not only violated the interests of her client, she also tarnished the reputation of the counseling profession and undoubtedly violated the rules of the counseling service for which she worked. Others who heard her story at the party may conclude that this is typical behavior for professional counselors in general or for those who work at that agency. As a result, they, too, may be reluctant to seek counseling for themselves or recommend it to others. This example represents a rather obvious breach of ethics that all responsible helping professionals would condemn, but other counselor behaviors are not so easily identified as ethical or unethical.

The following cases illustrate more complicated dilemmas that require more than good intentions and good sense to resolve ethically:

- A counselor's spouse's boss asks that the counselor see her college-aged son in counseling because he is confused and lonely. The spouse is being considered for a promotion. The counselor works at the college the son attends.
- The parents of a tenth-grader see a school counselor because they want her to counsel their son, who is dating a girl of another race. They say they're old-fashioned and believe that "people should stay with their own kind."
- A 79-year-old man sees a counselor because he wants to maintain his own home, but his children and neighbors insist that he can no longer care for it properly. He threatens suicide if he loses his home, but he admits to having some problems with independent living.
- A colleague at a counseling agency routinely uses diagnostic categories more severe than the client's actual difficulties in order to get the "most insurance reimbursement possible and give clients the full benefit of counseling."
- A counseling session with an adult client ends at dinnertime. This client came to the counselor because of difficulties in social assertiveness. He asks the counselor to join him for dinner at his parents' restaurant.

In each of these dilemmas, it is not easy to discern what really promotes the welfare of the client, the employer, or the profession. The counselor who "fudges" diagnoses for insurance purposes appears to be attempting to promote the welfare of her clients. But is she really? The person who counsels the boss's son may truly help him in his life choices. Should the personal connection with this young man stop the counseling from taking place? After all, the connection to this young man comes only through the counselor's spouse. Would accepting the client's dinner invitation facilitate trust in counseling and reward the client's social assertiveness, thereby helping the client, or would it jeopardize the counselor's objectivity and therefore risk harm to the client? This chapter sorts out these and other ethical dilemmas by discussing the ethical standards and principles that underlie the counseling profession and make counselors better equipped to act responsibly in such situations.

CODES OF PROFESSIONAL ETHICS

When most people think of professional ethics, they think of written codes of conduct that identify one set of behaviors that is expected of all professionals (e.g., keeping counseling records confidential) and another set of behaviors that is prohibited to all members (e.g., inaccurately advertising one's credentials). Codes of ethics also include a set of aspirational principles that professionals can use to guide their actions. One example of these aspirational statements comes from the "Ethical Principles" of the American Psychological Association (1992): "Psychologists accord appropriate respect to the fundamental rights, dignity and worth of all people" (p. 1599). Every helping profession has a code of ethics, and some have special codes of ethics for particular kinds of counseling, such as group counseling. You will find the codes of ethics of the American Counseling Association (formerly known as the American Association for Counseling and Development) and the American Psychological Association in Appendixes A and B of this book. Each member of these professional associations is expected to read the code thoroughly and to abide by its contents.

Becoming familiar with a code of ethics is one way a counselor-in-training can be sensitized to situations that may be ethically problematic. Sensitivity to ethics is important because no client, colleague, or supervisor ever announces in advance that he or she is about to present an ethical dilemma to the counselor. Usually, the other party is oblivious to the ethical dimensions of the situation. Thus, the counselor must be able to independently identify an ethical dilemma as soon as it arises. Ethics scholars argue that counselors who are ethically sensitive are less likely to inadvertently choose inappropriate alternatives. Please read the codes in Appendixes A and B before you continue with this chapter.

Note that a good deal of the content of the codes overlaps, although the organization of that content differs somewhat. Each code includes a discussion of how the helping professional ought to relate to clients, to colleagues, to employers, and to the public, emphasizing the responsibility to promote client welfare as the main goal. Each also includes statements describing the client's rights to privacy, accurate assessment of needs, freedom of choice, and fair and competent treatment or referral. Both codes define the responsibilities of helping professionals to present their training and experience honestly, to be direct and fair in assigning fees for services, to keep current with changes in practice, to conduct research responsibly, and to train competent future professionals. They elaborate on how professionals ought to balance their professional responsibilities to clients and to employers and explicitly prohibit certain behaviors such as sexual intimacies with clients, claiming titles or degrees not earned, and using psychological tests inappropriately. Each code recognizes the relationship between ethical standards and the law and advises practitioners to abide by the laws of their jurisdiction

unless blatantly unethical. These ethical standards are also fairly explicit about counselors' responsibilities to and for other counseling professionals. For example, each code addresses the responsibility of a counselor when he or she witnesses or learns of a colleague's unethical conduct. Although the wording differs somewhat, the essence is that each helping professional has an ethical responsibility to try to right any wrong he or she knows of and to see to it that the colleague works to correct the problem. In the case of a serious breach of ethics, the codes demand that the colleague's misbehavior be reported to the appropriate ethics committee. In other words, the ethics codes obligate counselors not only to act ethically in their own interactions, but also to intervene at some level in known ethical violations of colleagues.

Ethical standards not only sensitize a helping professional to the kinds of ethical issues likely to arise in practice, they also act as a first resource when such a professional is faced with a dilemma about what really promotes the client's well-being. To illustrate, when trying to decide whether to counsel the boss's son in the above example, the counselor can refer to the "Ethical Standards" of the American Association for Counseling and Development (1988); now the American Counseling Association, and read the following from Section B:

> When the member has other relationships, particularly of an administrative, supervisory, and or evaluative nature with an individual seeking counseling services, the member must not serve as counselor, but should refer the individual to another professional. . . . Dual relationships with clients that might impair the member's objectivity and professional judgment (e.g., as with close friends or relatives) must be avoided.

Although this section does not specifically identify the relatives of a spouse's boss as inappropriate clients, it does prohibit counseling persons with whom the counselor has another relationship, especially if that relationship might impair objectivity. The existence of two ways of relating to the client is called a *dual relationship.* The counselor has a dual relationship with the young man because he would be both a client and a relative of a significant person in the counselor's spouse's life. Another example of a dual relationship is if the same person were both a client and a student. This dual relationship is unethical because the counselor is in an evaluative relationship with the prospective client. As a teacher, that professional must evaluate the quality of the student's work. Not all dual relationships are automatically problematic. As the quote from the "Ethical Standards" indicates, dual relationships in which the objectivity or professional judgment of the counselor may be compromised are unethical. The careful reader of the code can then surmise that accepting the son of a spouse's boss as a client is unethical because the counselor may not be able to be objective about the client's needs. For instance, the counselor might hesitate to recommend a course of action to the client with which the boss is likely to disagree because of the possible repercussions on the spouse. The counselor in this case has another reason to see the client other than the client's well-being (i.e., impressing the boss and getting the spouse promoted). When self-interest becomes a prominent motivation, accepting a client becomes unethical. It is worthwhile to note that accepting this client may not really be in the best interests of the counselor either. The counseling may fail and the spouse would be worse off, or the counseling may succeed from the client's perspective but fail in the boss's view and the spouse may still suffer harm. The boss may also expect the counselor to reveal confidential information to her, putting the counselor at risk of betraying the client's trust.

Similarly, in the case of the counselor who reports more extreme diagnoses to insurance agencies in order to "beat the bureaucracy" and get what she sees as a full measure of reimbursement for needed counseling time, Section F of the "Ethical Standards" provides the following guidance: "A member must adhere to the regulations for private practice of the locality where the services are offered." Misrepresenting a diagnosis is prohibited because it is illegal and represents fraud. Therefore, such behavior by a counselor is unethical, regardless of its motivation. Although insurers sometimes have standards for care with which counselors disagree, disagreement about what is appropriate for the client does not justify lying to the insurance company. The counselor may ethically assist the client in appealing an insurance denial, may complain to the insurer about its rules, or ask the professional association to lobby the insurance industry and legislative bodies for more appropriate standards for care. However, the counselor who knowingly falsifies a diagnosis is subject to both ethical and legal penalties. Deceiving an insurance company about a diagnosis is also unethical in another respect. The private practice for which the counselor works probably has policies that prohibit such actions. The "Ethical Standards," Section A, state that the counselor has an ethical obligation to abide by the policies of the employer (as long as those policies are not obviously unethical). Thus, the counselor's actions clearly violate two separate sections of the code. In this case, a careful reading of the code gives the counselor the answer to this ethical dilemma.

To illustrate the usefulness of the code further, if Linda, the loquacious counselor in the initial example in this chapter, had examined either of these codes of ethics she would have known that sharing confidential information from clients in a social setting is also clearly prohibited. In other words, in a number of situations faced by counselors in their daily work, the codes provide the information necessary to make responsible ethical decisions. Thus, it is critical for all helping professionals to be thoroughly knowledgeable about the current code of conduct for their profession. Many licensing and certification standards include assessment of ethical knowledge for this very reason.

In at least three other types of situations, the codes are less clear or even mute. First, the changing nature of the helping professions limits the usefulness of codes. Counseling practice changes rapidly, and new intervention strategies are developed with great frequency. In this pluralistic society, counselors serve increasingly diverse populations of clients, whose needs, values, and beliefs may differ substantially from those of more traditional populations. New populations often require new counseling approaches. However, codes of ethics can never be fully current because the process of updating a code is cumbersome and professional associations usually take several years to revise any code. In the case of innovations in practice, only the general aspirational statements from the code may apply. Thus, counselors cannot rely on codes of ethics to clearly identify the ethical choice under such circumstances. One example of this limitation of written codes of ethics is evident in the issue of confidentiality with HIV-positive clients, those who have been infected with the AIDS virus. Should confidentiality be broken to inform sexual partners of the HIV status of this person? The current codes do not address this specific issue, and the responsible counselor needs to look beyond the code to resolve this dilemma. One resource, of course, is the growing body of literature in professional journals and professional conferences on this and other emerging ethical issues (e.g., Harding, Gray, & Neal, 1993; Melton, 1988). In essence, the literature suggests that counselors faced with this dilemma should balance the client's right to privacy with the danger to a specific individual. If a client who is HIV-positive is having unprotected sexual contact with a specific partner who has no knowledge of the person's HIV status and the client shows little concern about the well-being of the partner, then the responsibility to break confidentiality is strong. On the other hand, the rights of the client also need consideration, and the implications of breaking confidentiality should be weighed carefully. Breaking confidentiality is likely to terminate counseling, and the client who needed assistance from a professional may now refuse to engage in the counseling process with any counselor. In addition, the potential to get the client to change his or her behavior toward sexual partners through counseling is also gone, and the counselor will have no impact whatsoever over the client's sexual behavior in the future. In any case, the counselor's decision about confidentiality should not be based on irrational fears about AIDS or any generalization that HIV-positive clients automatically have no rights to privacy and confidentiality because of their health status. The central issue should be a known danger to an identifiable victim.

The second limitation of codes of ethics is that no such code can address every complicated situation a counselor will encounter. Codes of conduct are not "blueprints for action," but rather signposts that identify the most obvious and frequently occurring ethical problems. The best interests of the client may not be clear, and some good and some bad may happen no matter what choice the counselor makes. The case of the 79-year-old man illustrates such a dilemma. Supporting the man in what he wants may not be in his best interest in the long run. At the same time, if the family's desires take priority, the man may indeed become depressed and lose his motivation for living, an outcome that would also be clearly detrimental to him. (The next section of this chapter will provide a strategy for trying to resolve such a complicated issue.)

Finally, there is a third way in which codes of ethics are an incomplete resource. If a counselor is a member of more than one professional association, it is possible that the statements in the codes will conflict with each other. The counselor then has to choose between codes but is still responsible for adhering to the codes of any organization that he or she has joined.

Obviously, then, a code of ethics cannot be the counselor's only resource in sorting out ethical dilemmas. No written standards can ever substitute for the judgment of the individual practitioner. Wrestling with complicated ethical dilemmas is as much a part of the counseling profession as building trust or choosing intervention strategies. The next section describes the ethical principles underlying the codes, which are the counselor's next resource should the codes be inadequate to resolve a dilemma.

ETHICAL PRINCIPLES

There are five ethical principles identified by Kitchener (1984, 1986) as fundamental to counseling ethics. These principles are respect for autonomy, beneficence (doing good), nonmaleficence (avoiding harm), justice (or fairness), and fidelity (or promise keeping). Kitchener argues that these five principles provide the primary rationale for the contents of codes of ethics of helping professions. Ethics codes inform professionals as to what behaviors are ethical or unethical; ethical principles explain *why* behaviors have been so labeled. Kitchener states that counselors need to turn to the broader principles when written codes fail to provide clear guidance for action. Taken together, these principles encompass the primary values shared by the profession and, in essence, its reasons for existence. This section defines each ethical principle and demonstrates how it can be used in reasoning about ethical dilemmas.

Respect for Autonomy

The first ethical principle is respect for autonomy, an individual's right to self-determination. Individuals have a right to think as they wish, even if others disagree. They also have the right to act freely even if others don't like their choices. Autonomy must be respected, with two restrictions. First, the rights of the individual end where others' rights begin. In other words, a person has the freedom to act as long as others' rights are not limited.

Second, respect for autonomy assumes that individuals are capable of understanding the implications and consequences of the choices they make. This ability to comprehend the meaning of choices is called competence. Society gives little autonomy to children because, at their developmental level, they are unlikely to be able to understand the implications of their choices. Similarly, the autonomy of persons with severe organic brain damage or those who are in the midst of psychotic episodes and not oriented to reality can also be restricted because these individuals are at least temporarily unaware of the meaning and consequences of their choices.

Let us return to the example of the aging man presented at the beginning of the chapter to illustrate the role of respect for autonomy in resolving ethical dilemmas. The autonomy principle suggests that the older man has the freedom to choose where he wants to live if he understands the meaning of that choice. His choice need not be logical or even in his ultimate best interest as long as he understands its meaning. In this case then, the principle of autonomy obligates the counselor to first evaluate the man's competence to make autonomous choices. If the man is competent, then the counselor must respect his choice as a priority, even if the counselor doesn't approve of it. The issue is not whether the counselor should decide for him what is best, but rather how the counselor can help him to live the way he wishes with the least possible risk to him and help him resolve the conflicts with those worried about him. Counselors need to avoid acting paternalistically—that is, in the role of a parent who "knows better" than the adult client. This is sometimes difficult because client values conflict with the values of their counselors. Some potential conflicts are in the areas of women's rights, abortion rights, and social justice. A client may believe that wives should be submissive to their husbands in all things, and the counselor may believe in equality between marital partners. The principle of autonomy means that counselors cannot ethically impose their values on clients or use their influence to get clients to "see things the right way." Client beliefs and values can and do change during counseling. The issue is not whether the counselor helps the client explore beliefs that are troubling, but rather whose agenda it is to explore those beliefs. If the client freely chooses to focus on values and is free to come to an independent decision based on the exploration, the counselor can ethically explore these delicate subjects.

Beneficence

The second ethical principle—beneficence—is at the core of the profession. As members of a profession whose justification for existence is to "do good" for others, counselors have deeper ethical responsibilities than ordinary friends or confidants who receive no payment for their trust and support and do not purport to have any special training in helping. Counselors publicize themselves as "expert helpers," and people therefore come to see them precisely at those moments when they think the support of loved ones won't be sufficient to help or when they're desperate to get out of a situation they can't cope with. Counselors must do all they can to help. Moreover, counselors' status as paid professional helpers means that leaving clients at the end of counseling in the same place as they began is also inconsistent with counselors' role. Without commitment and skills to help their clients, counselors are guilty of a kind of false advertising of their services. Of course, there are times when counseling has negative outcomes. What is important in these cases is whether the counselor did all within his or her control to assist the client. Thus, one additional criterion for evaluating whether a particular course of action is ethical or not is to ask: "Is this course of action likely to benefit the client?"

Nonmaleficence

The responsibility not to make the client worse by intention, reckless action, or incompetence is the third ethical principle, also called the principle of nonmaleficence. This principle is also at the foundation of biomedical ethics. Some have argued that it is the most fundamental ethical principle guiding all human service professions. Precisely because counselors profess to be helping professionals, they have a duty not to make a client worse if this outcome is avoidable. In the past, scholars thought that counseling was not a risky activity; they admitted that it did not always help but suggested that it could not really hurt a client either. In other words, it was seen as a fairly innocuous activity. More recent evidence has overwhelmingly contradicted that perception. Counseling and therapy can be powerful tools and can be used to a client's significant advantage *or* disadvantage. Thus, the burden on the counselor to assess client problems accurately, choose counseling strategies wisely, and monitor the impact of counseling on each client is great. It is this ethical principle that underlies the statements in the ethics codes about practicing within the "limits of one's competence"—that is, dealing with client problems with which one has been trained and using counseling strategies with which one is skilled unless under supervision. This ethical principle is also at the core of the requirement that clients not be exposed to research or experimental treatments with high risk and little hope of real benefit. Thus, when evaluating whether a course of action is ethical, the counselor must also ask whether the client is at risk for harm. If the answer is yes, then alternative courses of action are usually more ethical.

Justice

The fourth ethical principle is justice, or fairness. Justice is a value at the core of democratic societies, and it demands that persons be treated equally. Judgment about counseling goals and strategies must be based on the individual characteristics of the client and not on discriminatory attitudes toward groups. Stereotyping and bias are unethical because they are unjust, regardless of whether the discriminatory attitudes are conscious or not. Justice demands that no person be given better or worse treatment in counseling based on his or her status in society. Thus, when evaluating whether an action is ethical or not, a counselor needs to ask whether that action is based on any factor other than the unique needs of the individual.

The principle of justice also helps guide the counselor in responding to the parents of a tenth-grader who don't want their son dating a girl of a different race, one of the examples in the beginning of this chapter. If the counselor learns that the only attribute of the young woman the parents find offensive is her race, then justice demands that the counselor refrain from involving himself in the situation in the way they want. Counselors cannot allow themselves to be put into the service of discrimination. The counselor may work with the family to foster communication and help resolve the conflict, but to agree to try to dissuade the boy from dating the girl on the basis of her race is unethical. Justice also demands that counselors display respectful and unbiased attitudes when counseling clients who are different in culture, background, lifestyle, or gender. This principle also requires counselors to use counseling strategies appropriate to the culture of the client. Two documents are especially helpful in assisting counselors in responding sensitively to diverse clients. One is entitled *Principles Concerning the Counseling and Psychotherapy of Women* (Fitzgerald & Nutt, 1986), and the other is named *Draft of Guidelines for Providers of Psychological Services to Ethnic, Linguistic and Culturally Diverse Populations* (American Psychological Association, 1991).

Fidelity

The fifth ethical principle is fidelity. One helpful way to think about the principle of fidelity is to use of the term *promise keeping.* Counseling professionals are taught attitudes and skills that help build the client's trust and encourage his or her self-disclosure. Promoting trust is the counselor's main goal in initial counseling sessions because self-disclosure and trust are critical to the success of the counseling process. Once trust has been placed in the counselor, he or she becomes a powerful person who can do harm to the client. Since the counselor engages in a set of actions designed for the sole purpose of promoting trust, it is particularly despicable when that trust is betrayed. When counseling begins, counselors implicitly promise not to divulge what a client tells them unless there is some overwhelming reason that is ultimately in the client's or society's best interests. Research suggests that most clients assume that everything they tell a counselor will be kept confidential (Miller & Thelen, 1986). However, confidentiality is limited in some circumstances, and it is important for counselors to explain those limits to clients before self-disclosure begins. For example, ethical and legal standards demand that confidentiality be broken if the counselor learns of ongoing child abuse or if the client is at great risk to harm self or others. Thus, before clients disclose such information, they need to know the consequences of their disclosure.

Another way to think of fidelity is to use the word *loyalty.* The principle of fidelity demands that counselors be loyal to clients, to employers, and to the profession. Thus, current ethical standards prohibit the abandonment of a client in the midst of counseling. Clearly, there are good reasons why a counselor may need to terminate counseling before the client is finished. However, the principle of fidelity requires the counselor must provide for an appropriate referral for that client in order to be faithful to the initial promise to provide help. The principle of fidelity is also the reason why the codes of ethics include statements about responsibilities to employers and fellow professionals. To be ethical, counselors must be faithful to the mission of their employer, unless that mission interferes with the best interests of the clients. After all, counselors accept salaries and other benefits from their employer with the implied promise to do what the employer expects. Similarly, counselors enjoy the benefits of their professional status and so they must be loyal to the profession as well. Thus, when trying to decide on the ethics of an action, the counselor must ask, "Is this choice in keeping with the promises I have made, either implied or explicit?"

ETHICAL THEORY

There are times when the five ethical principles are also insufficient to resolve an ethical dilemma. The principle of beneficence, for example, sometimes conflicts with the principle of autonomy. In the example of the aging man who wishes to live alone, such a conflict is apparent: what may be most helpful to the man may contradict what he wishes. How is such a conflict resolved? Many scholars say that the most fundamental ethical principle is avoiding harm and that one must examine the alternatives and decide on the basis of which one risks the least harm. It is still possible, however, that an ethical dilemma will remain unsolved because the greater harm is not apparent. In these situations, Kitchener (1984) suggests that even broader ethical theories need to be consulted. These theories are broad frameworks that ethicists use for examining good and evil; examples are utilitarianism (the greatest good for the greatest number) and the moral law tradition (the definition of certain rules for human behavior as absolute). These ethical theories act as the ultimate criterion in resolving a professional ethical dilemma. For a more detailed discussion of ethical theories, see Kitchener (1984).

There are other resources for counselors agonizing over an ethical dilemma. Over the last two decades, the literature on professional ethics has expanded rapidly. There are several published models of ethical decision making (Haas & Malouf, 1989; Kitchener, 1984), several casebooks that give counselors practice in resolving ethical issues (APA, 1987; Herlihy & Golden, 1990), and a large body of articles that examine a wide range of specific ethical issues, such as confidentiality in counseling children and adolescents (Gustafson & McNamara, 1987), the ethics of group therapy (Corey & Corey, 1992) counseling supervision (Vasquez, 1992), and ethical responsibilities of researchers and faculty (Stanley, Sieber, & Melton, 1987; Welfel, 1992). The References list at the end of this chapter is a good introduction to this body of scholarship. In addition, professional conferences often include panel discussions about ethics and presentations of research findings on professional ethics. Finally, professional associations have ethics committees that can advise members about troublesome ethical dilemmas. In short, no counselor needs to feel that ethical decisions must be made in isolation and without consultation. Abundant resources are available to assist counselors in the difficult ethical decision-making process.

THE RELATIONSHIP BETWEEN ETHICS AND THE LAW

Ethical standards are internal guidelines developed by a profession to govern the activities of its membership. In fact, the existence of enforceable ethical standards is what distinguishes a profession from a trade. In contrast, the state laws regarding counseling practice are determined by the statutes passed by legislatures and the case law determined by court rulings. These laws vary widely from state to state. All fifty states now have a statute regarding mandated reporting of child abuse, for example, but exactly what behaviors are defined as abusive, which professionals are covered by the statutes, and the penalties that exist for failure to report differ from state to state. Thus, counselors must educate themselves about the statutes within each state where they practice.

In addition, case law from litigation involving therapists has come to influence the current contents of ethics codes. The most famous example is the 1976 Tarasoff case from California. The court's ruling in this case established what is called the "duty to warn" a potential victim when a client makes a threat to cause harm to a particular person. In this case, the therapist was sued by the victim's parents for failing to notify the victim that the client had made serious threats against her life. The victim's family won the case with the judge arguing that the therapist's duty in this case went beyond notifying the police (as had been done) of the client's murderous intent. The court asserted that the duty of the therapist was to break confidentiality and warn the intended victim. The court held that "the protective privilege ends where the public peril begins" (*Tarasoff* v. *Regents of the University of California,* 551 P.2d 334, p. 347). Subsequent to that ruling, the ethics codes of counseling, psychology, and social work were revised to include a duty to warn.

Most often, counselors are defendants in court cases because they have been sued for professional negligence, also called malpractice. In recent years, civil suits against mental health professionals have multiplied, and counselors have become increasingly concerned about how to avoid malpractice claims. At conferences and workshops, presentations about avoiding malpractice are often filled to capacity. The primary message about avoiding malpractice is to abide by the ethics code of the profession (Swensen, 1993). Counseling is a risky profession, and even responsible and skilled professionals get sued for malpractice. Those who abide by their profession's ethics code, understand the ethical principles underlying the code, and keep current with the literature about emerging ethical issues are as well protected as one can be against such lawsuits.

COMMON ETHICAL VIOLATIONS BY MENTAL HEALTH PROFESSIONALS

Over the last decade, much information has been gathered regarding the ethical practice of mental health professionals. Research has been conducted, and licensing boards and ethics committees have published summaries of the complaints presented to them each year. Taken together, these data provide a fairly clear picture of the kinds of ethical difficulties most frequently encountered by helping professionals. Ethics committees deal overwhelmingly with one type of ethics violation: dual relationships with clients, especially sexual intimacies with current or former clients. This violation is particularly troubling because of the clear evidence of harm coming to clients who have been sexually exploited by their counselors (Bouhoutsos, et al., 1983). This kind of behavior has been termed a form of sexual assault and has been compared to incest in its devastating effects on clients. (In fact, sexual exploitation of a client carries criminal penalties in some states.) Even more disturbing is the evidence that those who are sexually intimate with clients tend to have a pattern of such behavior (Holroyd & Brodsky, 1977). The personal difficulties that brought these clients to counseling in the first place are not helped; in fact, they are often worsened. Moreover, such clients will be averse to getting the therapy they need because of their victimization by a counselor. So troubling are the consequences of sexual exploitation by mental health professionals that the American Psychological Association has published a brochure intended to help clients learn about their rights (Committee on Women in Psychology, 1989).

Since counselors are human, it is not surprising that they sometimes feel sexual attraction toward a client. Experiencing such attraction is not unethical in itself. What is unethical is acting upon it. The first step for a counselor who experiences sexual feelings for a client is to consult with a supervisor or colleague to discuss the case and decide on an appropriate course of action. The frank discussion about the attraction is often sufficient to help the counselor attend to the client's concerns. If the consultation does not refocus the counselor's attention, referring the client to another counselor is usually the wisest choice. If sexual attraction to clients happens persistently and frequently, the counselor should seek counseling to better understand the personal issues that may be provoking this reaction. Obviously, a counselor who is having persistent sexual thoughts about a client is likely to be distracted from the client's needs and concerns. Sometimes counselors have responded to their attraction to a client by terminating the counseling relationship, referring the client to another professional, and then beginning a social relationship with that former client. Such a "solution" to the problem of sexual attraction to a client is unethical. The fact that counseling has officially ended does not immediately change the client's perception of the counselor as a professional or diminish the problems that provoked that client to seek counseling in the first place. Counselors need to base their working life on the assumption that clients will never be a population from which they will obtain friends or lovers. The risk of harm to clients is too great in such situations.

Other breaches of ethics that commonly come to the attention of licensing boards include other forms of unprofessional conduct (such as using counseling sessions to discuss the counselor's problems), unethical billing practices, incompetent practice, fraudulent application for license, violations of confidentiality, misrepresentation of competence, and violations relating to reporting of child abuse (Pope & Vasquez, 1991). Pope and Vetter (1992) surveyed psychologists about the kinds of ethical issues they encountered in their work. Those who responded to the survey reported more ethical dilemmas about confidentiality than about any other issue. They expressed concerns about how and when to disclose confidential information about minors and how to respond when one client has several caregivers or the same professional has clients who know each other. In addition, they also described ethical dilemmas about the definition of dual relationships that are not sexual and about payment, insurance, and the like. It is interesting to note that although confidentiality was the issue mentioned most frequently by the respondents in this survey, confidentiality violations are seldom reported to ethics committees. A national survey of certified counselors revealed that the overwhelming majority saw sexual exploitation of clients and violations of informed consent, confidentiality, and voluntary participation in counseling as unethical. Counselors were more uncertain about the ethics of charging and collecting fees for counseling and the ethics of nonsexual dual relationships (Gibson & Pope, 1993).

Which counselors commit these ethical violations? No particular demographic characteristics have been associated with unethical practice.

Experience, gender, type of degree, and similar characteristics do not predict who will act unethically, with one exception. Complaints about dual relationships with clients, especially sexual intimacies with clients or former clients, have been largely made against male mental health professionals. The usual pattern is of an older male therapist and a younger female client. However, cases against women therapists have also been reported.

Some counselors blunder into unethical actions because "they just didn't think about the ethical issues" or they weren't familiar with the code of ethics of the profession. Given the needs to establish a trusting relationship, to assess the client's concerns, and to develop appropriate strategies to assist the client, it is not altogether surprising that counselors get distracted from otherwise obvious ethical issues. However, counselors must maintain their attention to ethical issues even while attending to therapeutic concerns if they are to merit the title of professional.

Other counselors act unethically because their primary motivation is self-interest or because they think codes of ethics are for professionals who are less experienced or gifted than they are. Such grandiose thinking has led to serious ethical violations and puts a counselor at risk for a pattern of unethical behavior with a number of clients. Although most helping professionals are dedicated to serving their clients, some seem to show interest in nothing but self-gratification.

Professionals sometimes act unethically because they are distracted by personal difficulties or are made especially needy by a personal crisis. A counselor who is lonely and sad after a difficult divorce may step over the boundaries of the counseling relationship because of those personal needs. Such cases have come before ethics committees and licensing boards on numerous occasions. The codes of ethics all address this kind of circumstance, putting the responsibility to be aware of personal limitations on the counselor. If unable to attend to the client's needs because of personal difficulties, the responsible counselor must not attempt counseling until the personal issues are resolved. The power of the counselor to do harm is too great to allow such a distortion of the counseling process to occur or continue. In this type of situation, counseling for the counselor is the best solution. Some professional associations are beginning to develop resources to assist such distressed professionals in recovering their equilibrium.

SUMMARY

A counselor is in a position of power and trust and has a duty to be respectful of that special status. Counselors who act irresponsibly cause real harm to their clients, their employers, and the reputation of the profession. The most fundamental ethical imperative is to act in the best interests of the client and avoid actions that risk harm to him or her.

Written codes of ethics provide statements of the specific ethical duties of professionals and the aspirational principles that are endorsed by the profession. These codes should not just be read; they should be digested so that professionals will know how to respond as soon as an ethical issue arises. Familiarity with a code of ethics also assists the counselor in recognizing an issue as an ethical problem. Some counselors act unethically because they failed to see the ethical dimensions of a situation until too late.

Codes of ethics cannot resolve every ethical dilemma. Some issues are too new to be included in a code, and some situations are too complicated to resolve by reference to any written standards. To deal with those situations, the counselor must understand the five major ethical principles underlying the code: autonomy, beneficence, avoidance of harm, justice, and fidelity. These principles can guide a counselor's ethical decision making when the relevant code does not provide clear direction. The responsibility for ethical behavior rests with the individual practitioner. No code of ethics can substitute for the judgment of the individual professional.

REFERENCES

American Association for Counseling and Development. (1988). *Ethical standards* (rev. ed.). Alexandria, VA: Author.

American Psychological Association. (1987). *Casebook on ethical principles of psychologists.* Washington, DC: Author.

American Psychological Association. (1991). *Draft of guidelines for providers of psychological services to ethnic, linguistic, and culturally diverse populations.* Washington, DC: Author.

American Psychological Association. (1992). Ethical principles of psychologists and code of conduct. *American Psychologist, 47,* 1597–1611.

Bouhoutsos, J. C., Holroyd, J., Lerman, H., Forer, B., & Greenberg, M. (1983). Sexual intimacy between psychotherapists and patients. *Professional Psychology: Research and Practice, 20,* 112–115.

Committee on Women in Psychology. (1989). If sex enters into the psychotherapy relationship. *Professional Psychology: Research and Practice, 20,* 112–115.

Corey, M., & Corey, G. (1992). *Groups: Process and practice* (4th ed.). Pacific Grove, CA: Brooks/Cole.

Fitzgerald, L. F., & Nutt, R. (1986). The Division 17 principles concerning the counseling/psychotherapy of women: Rationale and implementation. *The Counseling Psychologist, 14,* 180–216.

Gibson, W. T., & Pope, K. S. (1993). The ethics of counseling: A national survey of certified counselors. *Journal of Counseling and Development, 71,* 330–336.

Gustafson, K. E., & McNamara, J. R. (1987). Confidentiality with minor clients: Issues and guidelines for therapists. *Professional Psychology: Research and Practice, 17,* 111–114.

Haas, L. J., & Malouf, J. F. (1989). *Keeping up the good work: A practitioner's guide to mental health ethics.* Sarasota, FL: Professional Resource Exchange.

Harding, A. K., Gray, L. A., & Neal, M. (1993). Confidentiality with clients who have HIV: A review of ethical and legal guidelines and professional policies. *Journal of Counseling and Development, 71,* 297–305.

Herlihy, B., & Golden, L. (1990). *Ethical standards casebook* (4th ed.). Alexandria, VA: American Association for Counseling and Development.

Holroyd, J. C., & Brodsky, A. M. (1977). Psychologists' attitudes and practices regarding erotic and nonerotic physical contact with patients. *American Psychologist, 32,* 843–849.

Kitchener, K. S. (1984). Intuition, critical evaluation and ethical principles: The foundation for ethical decisions in counseling psychology. *The Counseling Psychologist, 12,* 43–55.

Kitchener, K. S. (1986). Teaching applied ethics in counselor education: An integration of psychological and philosophical analysis. *Journal of Counseling and Development, 64,* 306–310.

Melton, G. B. (1988). Ethical and legal issues in AIDS-related practice. *American Psychologist, 43,* 941–947.

Miller, D. J., & Thelen, M. H. (1986). Knowledge and beliefs about confidentiality in psychotherapy. *Professional Psychology: Research and Practice, 17,* 15–19.

Pope, K. S., & Vasquez, M. J. T. (1991). *Ethics in psychotherapy and counseling.* San Francisco: Jossey-Bass.

Pope, K. S., & Vetter, V. A. (1992). Ethical dilemmas encountered by the members of the American Psychological Association: A national survey. *American Psychologist, 47,* 397–411.

Stanley, B., Sieber, J. E., & Melton, G. B. (1987). Empirical studies of ethical issues in research. *American Psychologist, 42,* 735–741.

Swensen, L. (1993). *Psychology and the law for the helping professions.* Pacific Grove, CA: Brooks/Cole.

Tarasoff v. *Regents of University of California,* 551 P.2d 334 (Cal, 1976).

Vasquez, M. J. T. (1992). Psychologist as clinical supervisor: Promoting ethical practice. *Professional Psychology: Research and Practice, 23,* 196–202.

Welfel, E. R. (1992). Psychologist as ethics educator: Successes, failures and unanswered questions. *Professional Psychology: Research and Practice, 23,* 182–189.

APPENDIX A

ETHICAL STANDARDS

AMERICAN COUNSELING ASSOCIATION (FORMERLY AMERICAN ASSOCIATION FOR COUNSELING AND DEVELOPMENT)

PREAMBLE

The Association is an educational, scientific, and professional organization whose members are dedicated to the enhancement of the worth, dignity, potential, and uniqueness of each individual and thus to the service of society.

The Association recognizes that the role definitions and work settings of its members include a wide variety of academic disciplines, levels of academic preparation, and agency services. This diversity reflects the breadth of the Association's interest and influence. It also poses challenging complexities in efforts to set standards for the performance of members, desired requisite preparation or practice, and supporting social, legal, and ethical controls.

The specification of ethical standards enables the Association to clarify to present and future members and to those served by members the nature of ethical responsibilities held in common by its members.

The existence of such standards serves to stimulate greater concern by members for their own professional functioning and for the conduct of fellow professionals such as counselors, guidance and student personnel workers, and others in the helping professions. As the ethical code of the Association, this document establishes principles that define the ethical behavior of Association members. Additional ethical guidelines developed by the Association's Divisions for their specialty areas may further define a member's ethical behavior.

Section A: General

1. The member influences the development of the profession by contin-

uous efforts to improve professional practices, teaching, services, and research. Professional growth is continuous throughout the member's career and is exemplified by the development of a philosophy that explains why and how a member functions in the helping relationship. Members must gather data on their effectiveness and be guided by the findings. Members recognize the need for continuing education to ensure competent service.

2. The member has a responsibility both to the individual who is served and to the institution within which the service is performed to maintain high standards of professional conduct. The member strives to maintain the highest levels of professional services offered to the individuals to be served. The member also strives to assist the agency, organization, or institution in providing the highest caliber of professional services. The acceptance of employment in an institution implies that the member is in agreement with the general policies and principles of the institution. Therefore the professional activities of the member are also in accord with the objectives of the institution. If, despite concerted efforts, the member cannot reach agreement with the employer as to acceptable standards of conduct that allow for changes in institutional policy conducive to the positive growth and development of clients, then terminating the affiliation should be seriously considered.
3. Ethical behavior among professional associates, both members and nonmembers, must be expected at all times. When information is possessed that raises doubt as to the ethical behavior of professional colleagues, whether Association members or not, the member must take action to attempt to rectify such a condition. Such action shall use the institution's channels first and then use procedures established by the Association.
4. The member neither claims nor implies professional qualifications exceeding those possessed and is responsible for correcting any misrepresentations of these qualifications by others.
5. In establishing fees for professional counseling services, members must consider the financial status of clients and locality. In the event that the established fee structure is inappropriate for a client, assistance must be provided in finding comparable services of acceptable cost.
6. When members provide information to the public or to subordinates, peers, or supervisors, they have a responsibility to ensure that the content is general, unidentified client information that is accurate, unbiased, and consists of objective, factual data.
7. Members recognize their boundaries of competence and provide only those services and use only those techniques for which they are qualified by training or experience. Members should only accept those positions for which they are professionally qualified.
8. In the counseling relationship, the counselor is aware of the intimacy of the relationship and maintains respect for the client and avoids engaging in activities that seek to meet the counselor's personal needs at the expense of that client.
9. Members do not condone or engage in sexual harassment which is defined as deliberate or repeated comments, gestures, or physical contacts of a sexual nature.

10. The member avoids bringing personal issues into the counseling relationship, especially if the potential for harm is present. Through awareness of the negative impact of both racial and sexual stereotyping and discrimination, the counselor guards the individual rights and personal dignity of the client in the counseling relationship.
11. Products or services provided by the member by means of classroom instruction, public lectures, demonstrations, written articles, radio or television programs, or other types of media must meet the criteria cited in these Standards.

Section B: Counseling Relationship

This section refers to practices and procedures of individual and/or group counseling relationships.

The member must recognize the need for client freedom of choice. Under those circumstances where this is not possible, the member must apprise clients of restrictions that may limit their freedom of choice.

1. The member's primary obligation is to respect the integrity and promote the welfare of the client(s), whether the client(s) is (are) assisted individually or in a group relationship. In a group setting, the member is also responsible for taking reasonable precautions to protect individuals from physical and/or psychological trauma resulting from interaction within the group.
2. Members make provisions for maintaining confidentiality in the storage and disposal of records and follow an established record retention and disposition policy. The counseling relationship and information resulting therefrom must be kept confidential, consistent with the obligations of the member as a professional person. In a group counseling setting, the counselor must set a norm of confidentiality regarding all group participants' disclosures.
3. If an individual is already in a counseling relationship with another professional person, the member does not enter into a counseling relationship without first contacting and receiving the approval of that other professional. If the member discovers that the client is in another counseling relationship after the counseling relationship begins, the member must gain the consent of the other professional or terminate the relationship, unless the client elects to terminate the other relationship.
4. When the client's condition indicates that there is clear and imminent danger to the client or others, the member must take reasonable personal action or inform responsible authorities. Consultation with other professionals must be used where possible. The assumption of responsibility for the client's(s') behavior must be taken only after careful deliberation. The client must be involved in the resumption of responsibility as quickly as possible.
5. Records of the counseling relationship, including interview notes, test data, correspondence, tape recordings, electronic data storage, and other documents are to be considered professional information for use in counseling, and they should not be considered a part of the records of the institution or agency in which the counselor is employed unless specified by state statute or regulation. Revelation to others of counseling material must occur only

upon the expressed consent of the client.

6. In view of the extensive data storage and processing capacities of the computer, the member must ensure that data maintained on a computer is: (a) limited to information that is appropriate and necessary for the services being provided; (b) destroyed after it is determined that the information is no longer of any value in providing services; and (c) restricted in terms of access to appropriate staff members involved in the provision of services by using the best computer security methods available.
7. Use of data derived from a counseling relationship for purposes of counselor training or research shall be confined to content that can be disguised to ensure full protection of the identity of the subject client.
8. The member must inform the client of the purposes, goals, techniques, rules of procedure, and limitations that may affect the relationship at or before the time that the counseling relationship is entered. When working with minors or persons who are unable to give consent, the member protects these clients' best interests.
9. In view of common misconceptions related to the perceived inherent validity of computer generated data and narrative reports, the member must ensure that the client is provided with information as part of the counseling relationship that adequately explains the limitations of computer technology.
10. The member must screen prospective group participants, especially when the emphasis is on self-understanding and growth through self-disclosure. The member must maintain an awareness of the group participants' compatibility throughout the life of the group.
11. The member may choose to consult with any other professionally competent person about a client. In choosing a consultant, the member must avoid placing the consultant in a conflict of interest situation that would preclude the consultant's being a proper party to the member's efforts to help the client.
12. If the member determines an inability to be of professional assistance to the client, the member must either avoid initiating the counseling relationship or immediately terminate that relationship. In either event, the member must suggest appropriate alternatives. (The member must be knowledgeable about referral resources so that a satisfactory referral can be initiated.) In the event the client declines the suggested referral, the member is not obligated to continue the relationship.
13. When the member has other relationships, particularly of an administrative, supervisory, and/or evaluative nature with an individual seeking counseling services, the member must not serve as the counselor but should refer the individual to another professional. Only in instances where such an alternative is unavailable and where the individual's situation warrants counseling intervention should the member enter into and/or maintain a counseling relationship. Dual relationships with clients that might impair the member's objectivity and professional judgment (e.g., as with close friends or relatives) must be avoided and/or the counseling relationship terminated through referral to another competent professional.

14. The member will avoid any type of sexual intimacies with clients. Sexual relationships with clients are unethical.
15. All experimental methods of treatment must be clearly indicated to prospective recipients, and safety precautions are to be adhered to by the member.
16. When computer applications are used as a component of counseling services, the member must ensure that: (a) the client is intellectually, emotionally, and physically capable of using the computer application; (b) the computer application is appropriate for the needs of the client; (c) the client understands the purpose and operation of the computer application; and (d) that a follow-up of client use of a computer application is provided to both correct possible problems (misconceptions or inappropriate use) and assess subsequent needs.
17. When the member is engaged in short-term group treatment/training programs (e.g., marathons and other encounter-type or growth groups), the member ensures that there is professional assistance available during and following the group experience.
18. Should the member be engaged in a work setting that calls for any variation from the above statements, the member is obligated to consult with other professionals whenever possible to consider justifiable alternatives.
19. The member must ensure that members of various ethnic, racial, religious, disability, and socioeconomic groups have equal access to computer applications used to support counseling services and that the content of available computer applications does not discriminate against the groups described above.
20. When computer applications are developed by the member for use by the general public as self-help stand-alone computer software, the member must ensure that: (a) self-help computer applications are designed from the beginning to function in a stand-alone manner, as opposed to modifying software that was originally designed to require support from a counselor; (b) self-help computer applications will include within the program statements regarding intended user outcomes, suggestions for using the software, a description of the conditions under which self-help computer applications might not be appropriate, and a description of when and how counseling services might be beneficial; and (c) the manual for such applications will include the qualifications of the developer, the development process, validation data, and operating procedures.

Section C: Measurement and Evaluation

The primary purpose of educational and psychological testing is to provide descriptive measures that are objective and interpretable in either comparable or absolute terms. The member must recognize the need to interpret the statements that follow as applying to the whole range of appraisal techniques including test and nontest data. Test results constitute only one of a variety of pertinent sources of information for personnel, guidance, and counseling decisions.

1. The member must provide specific orientation or information to the examinee(s) prior to and following the test administration so that the

results of testing may be placed in proper perspective with other relevant factors. In so doing, the member must recognize the effects of socioeconomic, ethnic, and cultural factors on test scores. It is the member's professional responsibility to use additional unvalidated information carefully in modifying interpretation of the test results.

2. In selecting tests for use in a given situation or with a particular client, the member must consider carefully the specific validity, reliability, and appropriateness of the test(s). General validity, reliability, and related issues may be questioned legally as well as ethically when tests are used for vocational and educational selection, placement, or counseling.
3. When making any statements to the public about tests and testing, the member must give accurate information and avoid false claims or misconceptions. Special efforts are often required to avoid unwarranted connotations of such terms as IQ and grade equivalent scores.
4. Different tests demand different levels of competence for administration, scoring, and interpretation. Members must recognize the limits of their competence and perform only those functions for which they are prepared. In particular, members using computer-based test interpretations must be trained in the construct being measured and the specific instrument being used prior to using this type of computer application.
5. In situations where a computer is used for test administration and scoring, the member is responsible for ensuring that administration and scoring programs function properly to provide clients with accurate test results.
6. Tests must be administered under the same conditions that were established in their standardization. When tests are not administered under standard conditions or when unusual behavior or irregularities occur during the testing session, those conditions must be noted and the results designated as invalid or of questionable validity. Unsupervised or inadequately supervised test-taking, such as the use of tests through the mails, is considered unethical. On the other hand, the use of instruments that are so designed or standardized to be self-administered and self-scored, such as interest inventories, is to be encouraged.
7. The meaningfulness of test results used in personnel, guidance, and counseling functions generally depends on the examinee's unfamiliarity with the specific items on the test. Any prior coaching or dissemination of the test materials can invalidate test results. Therefore, test security is one of the professional obligations of the member. Conditions that produce most favorable test results must be made known to the examinee.
8. The purpose of testing and the explicit use of the results must be made known to the examinee prior to testing. The counselor must ensure that instrument limitations are not exceeded and that periodic review and/or retesting are made to prevent client stereotyping.
9. The examinee's welfare and explicit prior understanding must be the criteria for determining the recipients of the test results. The member must see that specific interpretation accompanies any release of individual or group test data. The interpretation of test data must be related to the examinee's particular concerns.

10. Members responsible for making decisions based on test results have an understanding of educational and psychological measurement, validation criteria, and test research.
11. The member must be cautious when interpreting the results of research instruments possessing insufficient technical data. The specific purposes for the use of such instruments must be stated explicitly to examinees.
12. The member must proceed with caution when attempting to evaluate and interpret the performance of minority group members or other persons who are not represented in the norm group on which the instrument was standardized.
13. When computer-based interpretations are developed by the member to support the assessment process, the member must ensure that the validity of such interpretations is established prior to the commercial distribution of such a computer application.
14. The member recognizes that test results may become obsolete. The member will avoid and prevent the misuse of obsolete test results.
15. The member must guard against the appropriation, reproduction, or modification of published tests or parts thereof without acknowledgment and permission from the previous publisher.

Section D: Research and Publication

1. Guidelines on research with human subjects shall be adhered to, such as:
   a. Ethical Principles in the Conduct of Research with Human Participants, Washington, D.C.: American Psychological Association, Inc., 1982.
   b. Code of Federal Regulations, Title 45, Subtitle A, Part 46, as currently issued.
   c. *Ethical Principles of Psychologists,* American Psychological Association, Principle #9: Research with Human Participants.
   d. Family Educational Rights and Privacy Act (the "Buckley Amendment").
   e. Current federal regulations and various state rights privacy acts.
2. In planning any research activity dealing with human subjects, the member must be aware of and responsive to all pertinent ethical principles and ensure that the research problem, design, and execution are in full compliance with them.
3. Responsibility for ethical research practice lies with the principal researcher, while others involved in the research activities share ethical obligation and full responsibility for their own actions.
4. In research with human subjects, researchers are responsible for the subjects' welfare throughout the experiment, and they must take all reasonable precautions to avoid causing injurious psychological, physical, or social effects on their subjects.
5. All research subjects must be informed of the purpose of the study except when withholding information or providing misinformation to them is essential to the investigation. In such research the member must be responsible for corrective action as soon as possible following completion of the research.
6. Participation in research must be voluntary. Involuntary participation is appropriate only when it can be

demonstrated that participation will have no harmful effects on subjects and is essential to the investigation.

7. When reporting research results, explicit mention must be made of all variables and conditions known to the investigator that might affect the outcome of the investigation or the interpretation of the data.
8. The member must be responsible for conducting and reporting investigations in a manner that minimizes the possibility that results will be misleading.
9. The member has an obligation to make available sufficient original research data to qualified others who may wish to replicate the study.
10. When supplying data, aiding in the research of another person, reporting research results, or in making original data available, due care must be taken to disguise the identity of the subjects in the absence of specific authorization from such subjects to do otherwise.
11. When conducting and reporting research, the member must be familiar with and give recognition to previous work on the topic, as well as to observe all copyright laws and follow the principles of giving full credit to all to whom credit is due.
12. The member must give due credit through joint authorship, acknowledgment, footnote statements, or other appropriate means to those who have contributed significantly to the research and/or publication, in accordance with such contributions.
13. The member must communicate to other members the results of any research judged to be of professional or scientific value. Results reflecting unfavorably on institutions, programs, services, or vested interests must not be withheld for such reasons.
14. If members agree to cooperate with another individual in research and/or publication, they incur an obligation to cooperate as promised in terms of punctuality of performance and with full regard to the completeness and accuracy of the information required.
15. Ethical practice requires that authors not submit the same manuscript or one essentially similar in content for simultaneous publication consideration by two or more journals. In addition, manuscripts published in whole or in substantial part in another journal or published work should not be submitted for publication without acknowledgment and permission from the previous publication.

Section E: Consulting

Consultation refers to a voluntary relationship between a professional helper and help-needing individual, group, or social unit in which the consultant is providing help to the client(s) in defining and solving a work-related problem or potential problem with a client or client system.

1. The member acting as a consultant must have a high degree of self-awareness of his/her own values, knowledge, skills, limitations, and needs in entering a helping relationship that involves human and/or organizational change and that the focus of the relationship be on the issues to be resolved and not on the person(s) presenting the problem.
2. There must be understanding and agreement between member and client for the problem definition, change of goals, and prediction of

consequences of interventions selected.

3. The member must be reasonably certain that she/he or the organization represented has the necessary competencies and resources for giving the kind of help that is needed now or may be needed later and that appropriate referral resources are available to the consultant.
4. The consulting relationship must be one in which client adaptability and growth toward self-direction are encouraged and cultivated. The member must maintain this role consistently and not become a decision maker for the client or create a future dependency on the consultant.
5. When announcing consultant availability for services, the member conscientiously adheres to the Association's Ethical Standards.
6. The member must refuse a private fee or other remuneration for consultation with persons who are entitled to these services through the member's employing institution or agency. The policies of a particular agency may make explicit provisions for private practice with agency clients by members of its staff. In such instances, the clients must be apprised of other options open to them should they seek private counseling services.

Section F: Private Practice

1. The member should assist the profession by facilitating the availability of counseling services in private as well as public settings.
2. In advertising services as a private practitioner, the member must advertise the services in a manner that accurately informs the public of professional services, expertise, and techniques of counseling available. A member who assumes an executive leadership role in the organization shall not permit his/her name to be used in professional notices during periods when he/she is not actively engaged in the private practice of counseling.
3. The member may list the following: highest relevant degree, type and level of certification and/or license, address, telephone number, office hours, type and/or description of services, and other relevant information. Such information must not contain false, inaccurate, misleading, partial, out-of-context, or deceptive material or statements.
4. Members do not present their affiliation with any organization in such a way that would imply inaccurate sponsorship or certification by that organization.
5. Members may join in partnership/corporation with other members and/or other professionals provided that each member of the partnership or corporation makes clear the separate specialties by name in compliance with the regulations of the locality.
6. A member has an obligation to withdraw from a counseling relationship if it is believed that employment will result in violation of the Ethical Standards. If the mental or physical condition of the member renders it difficult to carry out an effective professional relationship or if the member is discharged by the client because the counseling relationship is no longer productive for the client, then the member is obligated to terminate the counseling relationship.
7. A member must adhere to the regulations for private practice of the locality where the services are offered.
8. It is unethical to use one's institutional affiliation to recruit clients for one's private practice.

Section G: Personnel Administration

It is recognized that most members are employed in public or quasi-public institutions. The functioning of a member within an institution must contribute to the goals of the institution and vice versa if either is to accomplish their respective goals or objectives. It is therefore essential that the member and the institution function in ways to: (a) make the institution's goals explicit and public; (b) make the member's contribution to institutional goals specific; and (c) foster mutual accountability for goal achievement.

To accomplish these objectives, it is recognized that the member and the employer must share responsibilities in the formulation and implementation of personnel policies.

1. Members must define and describe the parameters and levels of their professional competency.
2. Members must establish interpersonal relations and working agreements with supervisors and subordinates regarding counseling or clinical relationships, confidentiality, distinction between public and private material, maintenance and dissemination of recorded information, work load, and accountability. Working agreements in each instance must be specified and made known to those concerned.
3. Members must alert their employers to conditions that may be potentially disruptive or damaging.
4. Members must inform employers of conditions that may limit their effectiveness.
5. Members must submit regularly to professional review and evaluation.
6. Members must be responsible for inservice development of self and/or staff.
7. Members must inform their staff of goals and programs.
8. Members must provide personnel practices that guarantee and enhance the rights and welfare of each recipient of their service.
9. Members must select competent persons and assign responsibilities compatible with their skills and experiences.
10. The member, at the onset of a counseling relationship, will inform the client of the member's intended use of supervisors regarding the disclosure of information concerning this case. The member will clearly inform the client of the limits of confidentiality in the relationship.
11. Members, as either employers or employees, do not engage in or condone practices that are inhumane, illegal, or unjustifiable (such as considerations based on sex, handicap, age, race) in hiring, promotion, or training.

Section H: Preparation Standards

Members who are responsible for training others must be guided by the preparation standards of the Association and relevant Division(s). The member who functions in the capacity of trainer assumes unique ethical responsibilities that frequently go beyond that of the member who does not function in a training capacity. These ethical responsibilities are outlined as follows:

1. Members must orient students to program expectations, basic skills development, and employment prospects prior to admission to the program.
2. Members in charge of learning experiences must establish programs that integrate academic study and supervised practice.

3. Members must establish a program directed toward developing students' skills, knowledge, and self-understanding, stated whenever possible in competency or performance terms.
4. Members must identify the levels of competencies of their students in compliance with relevant Division standards. These competencies must accommodate the paraprofessional as well as the professional.
5. Members, through continual student evaluation and appraisal, must be aware of the personal limitations of the learner that might impede future performance. The instructor must not only assist the learner in securing remedial assistance but also screen from the program those individuals who are unable to provide competent services.
6. Members must provide a program that includes training in research commensurate with levels of role functioning. Paraprofessional and technician-level personnel must be trained as consumers of research. In addition, personnel must learn how to evaluate their own and their program's effectiveness. Graduate training, especially at the doctoral level, would include preparation for original research by the member.
7. Members must make students aware of the ethical responsibilities and standards of the profession.
8. Preparatory programs must encourage students to value the ideals of service to individuals and to society. In this regard, direct financial remuneration or lack thereof must not influence the quality of service rendered. Moneary considerations must not be allowed to overshadow professional and humanitarian needs.
9. Members responsible for educational programs must be skilled as teachers and practitioners.
10. Members must present thoroughly varied theoretical positions so that students may make comparisons and have the opportunity to select a position.
11. Members must develop clear policies within their educational institutions regarding field placement and the roles of the student and the instructor in such placement.
12. Members must ensure that forms of learning focusing on self-understanding or growth are voluntary, or if required as part of the educational program, are made known to prospective students prior to entering the program. When the educational program offers a growth experience with an emphasis on self-disclosure or other relatively intimate or personal involvement, the member must have no administrative, supervisory, or evaluating authority regarding the participant.
13. The member will at all times provide students with clear and equally acceptable alternatives for self-understanding or growth experiences. The member will assure students that they have a right to accept these alternatives without prejudice or penalty.
14. Members must conduct an educational program in keeping with the current relevant guidelines of the Association.

APPENDIX B

ETHICAL PRINCIPLES OF PSYCHOLOGISTS

AMERICAN PSYCHOLOGICAL ASSOCIATION

INTRODUCTION

The American Psychological Association's (APA's) Ethical Principles of Psychologists and Code of Conduct (hereinafter referred to as the Ethics Code) consists of an Introduction, a Preamble, six General Principles (A-F), and specific Ethical Standards. The Introduction discusses the intent, organization, procedural considerations, and scope of application of the Ethics Code. The Preamble and General Principles are *aspirational* goals to guide psychologists toward the highest ideals of psychology. Although the Preamble and General Principles are not themselves enforceable rules, they should be considered by psychologists in arriving at an ethical course of action and may be considered by ethics bodies in interpreting the Ethical Standards. The Ethical Standards set forth *enforceable* rules for conduct as psychologists. Most of the Ethical Standards are written broadly, in order to apply to psychologists in varied roles, although the application of an Ethical Standard may vary depending on the context. The Ethical Standards are not exhaustive. The fact that a given conduct is not specifically addressed by the Ethics Code does not mean that it is necessarily either ethical or unethical.

Membership in the APA commits members to adhere to the APA Ethics Code and to the rules and procedures used to implement it. Psychologists and students, whether or not they are APA members, should be aware that the Ethics Code may be applied to them by state psychology boards, courts, or other public bodies.

This Ethics Code applies only to psychologists' work-related activities, that is, activities that are part of the psychologists' scientific and professional functions or that are psychological in nature. It includes the clinical or counseling practice of psychology, research, teaching, supervision of trainees, development of assessment instruments, conducting assessments, educational

 The Ethical Principles and Code of Conduct apply to psychologists, to students of psychology, and to others who do work of a psychological nature under the supervision of a psychologist. They are intended for the guidance of nonmembers of the Association who are engaged in psychological research or practice.

counseling, organizational consulting, social intervention, administration, and other activities as well. These work-related activities can be distinguished from the purely private conduct of a psychologist, which ordinarily is not within the purview of the Ethics Code.

The Ethics Code is intended to provide standards of professional conduct that can be applied by the APA and by other bodies that choose to adopt them. Whether or not a psychologist has violated the Ethics Code does not by itself determine whether he or she is legally liable in a court action, whether a contract is enforceable, or whether other legal consequences occur. These results are based on legal rather than ethical rules. However, compliance with or violation of the Ethics Code may be admissible as evidence in some legal proceedings, depending on the circumstances.

In the process of making decisions regarding their professional behavior, psychologists must consider this Ethics Code, in addition to applicable laws and psychology board regulations. If the Ethics Code establishes a higher standard of conduct than is required by law, psychologist must meet the higher ethical standard. If the Ethics Code standard appears to conflict with the requirements of law, then psychologists make known their commitment to the Ethics Code and take steps to resolve the conflict in a responsible manner. If neither law nor the Ethics Code resolves an issue, psychologists should consider other professional materials[1] and the dictates of their own conscience, as well as seek consultation with others within the field when this is practical.

The procedures for filing, investigating, and resolving complaints of unethical conduct are described in the current Rules and Procedures of the APA Ethics Committee. The actions that APA may take for violations of the Ethics Code include actions such as reprimand, censure, termination of APA membership, and referral of the matter to other bodies. Complainants who seek remedies such as monetary damages in alleging ethical violations by a psychologist must resort to private negotiation, administrative bodies, or the courts. Actions that violate the Ethics Code may lead to the imposition of sanctions on a psychologist by bodies other than APA, including state psychological associations, other professional groups, psychology boards, other state or federal agencies, and payors for health services. In addition to actions for violation of the Ethics Code, the APA By-laws provide that APA may take action against a member after his or her conviction of a felony, expulsion or suspension from an affiliated state psychological association, or suspension or loss of licensure.

[1]Professional materials that are most helpful in this regard are guidelines and standards that have been adopted or endorsed by professional psychological organizations. Such guidelines and standards, whether adopted by the American Psychological Association (APA) or its Divisions, are not enforceable as such by this Ethics Code, but are of educative value to psychologists, courts, and professional bodies. Such materials include, but are not limited to, the APA's *General Guidelines for Providers of Psychological Services* (1987), *Specialty Guidelines for the Delivery of Services by Clinical Psychologists, Counseling Psychologists, Industrial/Organizational Psychologists, and School Psychologists* (1981), *Guidelines for Computer Based Tests and Interpretations* (1987), *Standards for Educational and Psychological Testing* (1985), *Ethical Principles in the Conduct of Research with Human Participants* (1982), *Guidelines for Ethical Conduct in the Care and Use of Animals* (1985), *Guidelines for Providers of Psychological Services to Ethnic, Linguistic, and Culturally Diverse Populations* (1990), *and Publication Manual of the American Psychological Association* (3rd ed., 1983). Materials not adopted by APA as a whole include the APA Division 41 (Forensic Psychology)/American Psychology–Law Society's *Specialty Guidelines for Forensic Psychologists* (1991).

PREAMBLE

Psychologists work to develop a valid and reliable body of scientific knowledge based on research. They may apply that knowledge to human behavior in a variety of contexts. In doing so, they perform many roles, such as researcher, educator, diagnostician, therapist, supervisor, consultant, administrator, social interventionist, and expert witness. Their goal is to broaden knowledge of behavior and, where appropriate, to apply it pragmatically to improve the condition of both the individual and society. Psychologists respect the central importance of freedom of inquiry and expression in research, teaching, and publication. They also strive to help the public in developing informed judgments and choices concerning human behavior. This Ethics Code provides a common set of values upon which psychologists build their professional and scientific work.

This Code is intended to provide both the general principles and the decision rules to cover most situations encountered by psychologists. It has as its primary goal the welfare and protection of the individuals and groups with whom psychologists work. It is the individual responsibility of each psychologist to aspire to the highest possible standards of conduct. Psychologists respect and protect human and civil rights, and do not knowingly participate in or condone unfair discriminatory practices.

The development of a dynamic set of ethical standards for a psychologist's work-related conduct requires a personal commitment to a lifelong effort to act ethically; to encourage ethical behavior by students, supervisees, employees, and colleagues, as appropriate; and to consult with others, as needed, concerning ethical problems. Each psychologist supplements, but does not violate, the Ethics Code's values and rules on the basis of guidance drawn from personal values, culture, and experience.

GENERAL PRINCIPLES

Principle A: Competence

Psychologists strive to maintain high standards of competence in their work. They recognize the boundaries of their particular competencies and the limitations of their expertise. They provide only those services and use only those techniques for which they are qualified by education, training, or experience. Psychologists are cognizant of the fact that the competencies required in serving, teaching, and/or studying groups of people vary with the distinctive characteristics of those groups. In those areas in which recognized professional standards do not yet exist, psychologists exercise careful judgment and take appropriate precautions to protect the welfare of those with whom they work. They maintain knowledge of relevant scientific and professional information related to the services they render, and they recognize the need for ongoing education. Psychologists make appropriate use of scientific, professional, technical, and administrative resources.

Principle B: Integrity

Psychologists seek to promote integrity in the science, teaching, and practice of psychology. In these activities psychologists are honest, fair, and respectful of others. In describing or reporting their qualifications, services, products, fees, research, or teaching, they do not make statements that are false, misleading, or deceptive. Psychologists strive to be aware of their own belief systems, values, needs, and limitations and the effect of these on their work. To the extent feasible, they attempt to clarify for relevant parties the roles they are performing and to function appropriately in accordance with those roles. Psychologists avoid improper and potentially harmful dual relationships.

Principle C: Professional and Scientific Responsibility

Psychologists uphold professional standards of conduct, clarify their professional roles and obligations, accept appropriate responsibility for their behavior, and adapt their methods to the needs of different populations. Psychologists consult with, refer to, or cooperate with other professionals and institutions to the extent needed to serve the best interests of their patients, clients, or other recipients of their services. Psychologists' moral standards and conduct are personal matters to the same degree as is true for any other person, except as psychologists' conduct may compromise their professional responsibilities or reduce the public's trust in psychology and psychologists. Psychologists are concerned about the ethical compliance of their colleagues' scientific and professional conduct. When appropriate, they consult with colleagues in order to prevent or avoid unethical conduct.

Principle D: Respect for People's Rights and Dignity

Psychologists accord appropriate respect to the fundamental rights, dignity, and worth of all people. They respect the rights of individuals to privacy, confidentiality, self-determination, and autonomy, mindful that legal and other obligations may lead to inconsistency and conflict with the exercise of these rights. Psychologists are aware of cultural, individual, and role differences, including those due to age, gender, race, ethnicity, national origin, religion, sexual orientation, disability, language, and socioeconomic status. Psychologists try to eliminate the effect on their work of biases based on those factors, and they do not knowingly participate in or condone unfair discriminatory practices.

Principle E: Concern for Others' Welfare

Psychologists seek to contribute to the welfare of those with whom they interact professionally. In their professional actions, psychologists weigh the welfare and rights of their patients or clients, students, supervisees, human research participants, and other affected persons, and the welfare of animal subjects of research. When conflicts occur among psychologists' obligations or concerns, they attempt to resolve these conflicts and to perform their roles in a responsible fashion that avoids or minimizes harm. Psychologists are sensitive to real and ascribed differences in power between themselves and others, and they do not exploit or mislead other people during or after professional relationships.

Principle F: Social Responsibility

Psychologists are aware of their professional and scientific responsibilities to the community and the society in which they work and live. They apply and make public their knowledge of psychology in order to contribute to human welfare. Psychologists are concerned about and work to mitigate the causes of human suffering. When undertaking research, they strive to advance human welfare and the science of psychology. Psychologists try to avoid misuse of their work. Psychologists comply with the law and encourage the development of law and social policy that serve the interests of their patients and clients and the public. They are encouraged to contribute a portion of their professional time for little or no personal advantage.

ETHICAL STANDARDS

1. General Standards

These General Standards are potentially applicable to the professional and scientific activities of all psychologists.

1.01 Applicability of the Ethics Code

The activity of a psychologist subject to the Ethics Code may be reviewed under these Ethical Standards only if the activity is part of his or her work-related functions or the activity is psychological in nature. Personal activities having no connection to or effect on psychological roles are not subject to the Ethics Code.

1.02 Relationship of Ethics and Law

If psychologists' ethical responsibilities conflict with law, psychologists make known their commitment to the Ethics Code and take steps to resolve the conflict in a responsible manner.

1.03 Professional and Scientific Relationship

Psychologists provide diagnostic, therapeutic, teaching, research, supervisory, consultative, or other psychological services only in the context of a defined professional or scientific relationship or role. (See also Standards 2.01, Evaluation, Diagnosis, and Interventions in Professional Context, and 7.02, Forensic Assessments.)

1.04 Boundaries of Competence

(a) Psychologists provide services, teach, and conduct research only within the boundaries of their competence, based on their education, training, supervised experience, or appropriate professional experience.

(b) Psychologists provide services, teach, or conduct research in new areas or involving new techniques only after first undertaking appropriate study, training, supervision, and/or consultation from persons who are competent in those areas or techniques.

(c) In those emerging areas in which generally recognized standards for preparatory training do not yet exist, psychologists nevertheless take reasonable steps to ensure the competence of their work and to protect patients, clients, students, research participants, and others from harm.

1.05 Maintaining Expertise

Psychologists who engage in assessment, therapy, teaching, research, organizational consulting, or other professional activities maintain a reasonable level of awareness of current scientific and professional information in their fields of activity, and undertake ongoing efforts to maintain competence in the skills they use.

1.06 Basis for Scientific and Professional Judgments

Psychologists rely on scientifically and professionally derived knowledge when making scientific or professional judgments or when engaging in scholarly or professional endeavors.

1.07 Describing the Nature and Results of Psychological Services

(a) When psychologists provide assessment, evaluation, treatment, counseling, supervision, teaching, consultation, research, or other psychological services to an individual, a group, or an organization, they provide, using language that is reasonably understandable to the recipient of those services, appropriate information beforehand about the nature of such services and appropriate information later about results and conclusions. (See also Standard 2.09, Explaining Assessment Results.)

(b) If psychologists will be precluded by law or by organizational roles from providing such information to particular individuals or groups, they so inform those individuals or groups at the outset of the service.

1.08 Human Differences

Where differences of age, gender, race, ethnicity, national origin, religion, sexual orientation, disability, language, or socioeconomic status significantly affect psychologists' work concerning particular individuals or groups, psychologists obtain the training, experience, consultation, or supervision necessary to ensure the competence of their services, or they make appropriate referrals.

1.09 Respecting Others

In their work-related activities, psychologists respect the rights of others to hold values, attitudes, and opinions that differ from their own.

1.10 Nondiscrimination

In their work-related activities, psychologists do not engage in unfair discrimination based on age, gender, race, ethnicity, national origin, religion, sexual orientation, disability, socioeconomic status, or any basis proscribed by law.

1.11 Sexual Harassment

(a) Psychologists do not engage in sexual harassment. Sexual harassment is sexual solicitation, physical advances, or verbal or nonverbal conduct that is sexual in nature, that occurs in connection with the psychologist's activities or roles as a psychologist, and that either: (1) is unwelcome, is offensive, or creates a hostile workplace environment, and the psychologist knows or is told this; or (2) is sufficiently severe or intense to be abusive to a reasonable person in the context. Sexual harassment can consist of a single intense or severe act or of multiple persistent or pervasive acts.

(b) Psychologists accord sexual-harassment complainants and respondents dignity and respect. Psychologists do not participate in denying a person academic admittance or advancement, employment, tenure, or promotion, based solely upon their having made, or their being the subject of, sexual harassment charges. This does not preclude taking action based upon the outcome of such proceedings or consideration of other appropriate information.

1.12 Other Harassment

Psychologists do not knowingly engage in behavior that is harassing or demeaning to persons with whom they interact in their work based on factors such as those persons' age, gender, race, ethnicity, national origin, religion, sexual orientation, disability, language, or socioeconomic status.

1.13 Personal Problems and Conflicts

(a) Psychologists recognize that their personal problems and conflicts may interfere with their effectiveness. Accordingly, they refrain from undertaking an activity when they know or should know that their personal problems are likely to lead to harm to a patient, client, colleague, student, research participant, or other person to whom they may owe a professional or scientific obligation.

(b) In addition, psychologists have an obligation to be alert to signs of, and to obtain assistance for, their personal problems at an early stage, in order to prevent significantly impaired performance.

(c) When psychologists become aware of personal problems that may interfere with their performing work-related duties adequately, they take appropriate measures, such as obtaining professional consultation or assistance, and determine whether they should limit, suspend, or terminate their work-related duties.

1.14 Avoiding Harm

Psychologists take reasonable steps to avoid harming their patients or clients, research participants, students, and others with whom they work, and to minimize harm where it is forseeable and unavoidable.

1.15 Misuse of Psychologists' Influence

Because psychologists' scientific and professional judgments and actions may affect the lives of others, they are alert to and guard against personal, financial, social, organizational, or political factors that might lead to misuse of their influence.

1.16 Misuse of Psychologists' Work

(a) Psychologists do not participate in activities in which it appears likely that their skills or data will be misused by others, unless corrective mechanisms are available. (See also Standard 7.04, Truthfulness and Candor.)

(b) If psychologists learn of misuse or misrepresentation of their work, they take reasonable steps to correct or minimize the misuse or misrepresentation.

1.17 Multiple Relationships

(a) In many communities and situations, it may not be feasible or reasonable for psychologists to avoid social or other nonprofessional contacts with persons such as patients, clients, students, supervisees, or research participants. Psychologists must always be sensitive to the potential harmful effects of other contacts on their work and on those persons with whom they deal. A psychologist refrains from entering into or promising another personal, scientific, professional, financial, or other relationship with such persons if it appears likely that such a relationship reasonably might impair the psychologists's objectivity or otherwise interfere with the psychologist's effectively performing his or her functions as a psychologist, or might harm or exploit the other party.

(b) Likewise, whenever feasible, a psychologist refrains from taking on professional or scientific obligations when preexisting relationships would create a risk of such harm.

(c) If a psychologist finds that, due to unforeseen factors, a potentially harmful multiple relationship has arisen, the psychologist attempts to resolve it with due regard for the best interests of the affected person and maximal compliance with the Ethics Code.

1.18 Barter (with Patients or Clients)

Psychologists ordinarily refrain from accepting goods, services, or other nonmonetary remuneration from patients or clients in return for psychological services because such arrangements create inherent potential for conflicts, exploitation, and distortion of the professional relationship. A psychologist may participate in bartering *only* if (1) it is not clinically contraindicated, *and* (2) the relationship is not exploitative. (See also Standards 1.17, Multiple Relationships, and 1.25, Fees and Financial Arrangements.)

1.19 Exploitative Relationships

(a) Psychologists do not exploit persons over whom they have supervisory, evaluative, or other authority such as students, supervisees, employees, research participants, and clients or patients. (See also Standards 4.05–4.07 regarding sexual involvement with clients or patients.)

(b) Psychologists do not engage in sexual relationships with students or supervisees in training over whom the psychologist has evaluative or direct authority, because such relationships are so likely to impair judgment or be exploitative.

1.20 Consultations and Referrals

(a) Psychologists arrange for appropriate consultations and referrals based principally on the best interests of their patients or clients, with appropriate consent, and subject to other relevant considerations, including applicable law and contractual obligations. (See also Standards 5.01, Discussing the Limits of Confidentiality, and 5.06, Consultations.)

(b) When indicated and professionally appropriate, psychologists cooperate with other professionals in order to serve their patients or clients effectively and appropriately.

(c) Psychologists' referral practices are consistent with law.

1.21 Third-Party Requests for Services

(a) When a psychologist agrees to provide services to a person or entity at the request of a third party, the psychologist clarifies to the extent feasible, at the outset of the service, the nature of the relationship with each party. This clarification includes the role of the psychologist (such as therapist, organizational consultant, diagnostician, or expert witness), the probable uses of the services provided or the information obtained, and the fact that there may be limits to confidentiality.

(b) If there is a forseeable risk of the psychologist's being called upon to perform conflicting roles because of the involvement of a third party, the psychologist clarifies the nature and direction of his or her responsibilities, keeps all parties appropriately informed as matters develop, and resolves the situation in accordance with this Ethics Code.

1.22 Delegation to and Supervision of Subordinates

(a) Psychologists delegate to their employees, supervisees, and research assistants only those responsibilities that such persons can reasonably be expected to perform competently, on the basis of their education, training, or experience, either independently or with the level of supervision being provided.

(b) Psychologists provide proper training and supervision to their employees or supervisees and take reasonable steps to see that such persons perform services responsibly, competently, and ethically.

(c) If institutional policies, procedures, or practices prevent fulfillment of this obligation, psychologists attempt to modify their role or to correct the situation to the extent feasible.

1.23 Documentation of Professional and Scientific Work

(a) Psychologists appropriately document their professional and scientific work in order to facilitate provision of services later by them or by other professionals, to ensure accountability, and to meet other requirements of institutions or the law.

(b) When psychologists have reason to believe that records of their professional services will be used in legal proceedings involving recipients of or participants in their work, they have a responsibility to create and maintain documentation in the kind of detail and quality that would be consistent with reasonable scrutiny in an adjudicative forum. (See also Standard 7.01, Professionalism, under Forensic Activities.)

1.24 Records and Data

Psychologists create, maintain, disseminate, store, retain, and dispose of records and data relating to their research, practice, and other work in accordance with law and in a manner that permits compliance with the requirements of this Ethics Code. (See also Standard 5.04, Maintenance of Records.)

1.25 Fees and Financial Arrangements

(a) As early as is feasible in a professional or scientific relationship, the psychologist and the patient, client, or other appropriate recipient of psychological services reach an agreement specifying the compensation and the billing arrangements.

(b) Psychologists do not exploit recipients of services or payors with respect to fees.

(c) Psychologists' fee practices are consistent with law.

(d) Psychologists do not misrepresent their fees.

(e) If limitations to services can be anticipated because of limitations in financing, this is discussed with the patient, client, or other appropriate recipient of services as early as is feasible. (See also Standard 4.08, Interruption of Services.)

(f) If the patient, client, or other recipient of services does not pay for services as agreed, and if the psychologist wishes to use collection agencies or legal measures to collect the fees, the psychologist first informs the person that such measures will be taken and provides that person an opportunity to make prompt payment. (See also Standard 5.11, Withholding Records for Nonpayment.)

1.26 Accuracy in Reports to Payors and Funding Sources

In their reports to payors for services or sources of research funding, psychologists accurately state the nature of the research or service provided, the fees or charges, and where applicable, the identity of the provider, the findings, and the diagnosis. (See also Standard 5.05, Disclosures.)

1.27 Referrals and Fees

When a psychologist pays, receives payment from, or divides fees with another professional other than in an employer-employee relationship, the payment to each is based on the services (clinical, consultative, administrative, or other) provided and is not based on the referral itself.

2. EVALUATION, ASSESSMENT, OR INTERVENTION

2.01 Evaluation, Diagnosis, and Interventions in Professional Context

(a) Psychologists perform evaluations, diagnostic services, or interventions only within the context of a defined professional relationship. (See also Standard 1.03, Professional and Scientific Relationship.)

(b) Psychologists' assessments, recommendations, reports, and psychological diagnostic or evaluative statements are based on information and techniques (including personal interviews of the individual when appropriate) sufficient to provide appropriate substantiation for their findings. (See also Standard 7.02, Forensic Assessments.)

2.02 Competence and Appropriate Use of Assessments and Interventions

(a) Psychologists who develop, administer, score, interpret, or use psychological assessment techniques, interviews, tests, or instruments do so in a manner and for purposes that are appropriate in light of the research on or evidence of the usefulness and proper application of the techniques.

(b) Psychologists refrain from misuse of assessment techniques, interventions, results, and interpretations and take reasonable steps to prevent others from misusing the information these techniques provide. This includes refraining from releasing raw test results or raw data to persons, other than to patients or clients as appropriate, who are not qualified to use such information. (See also Standards 1.02, Relationship of Ethics and Law, and 1.04, Boundaries of Competence.)

2.03 Test Construction

Psychologists who develop and conduct research with tests and other assessment techniques use scientific procedures and current professional knowledge for test design, standardization, validation, reduction or elimination of bias, and recommendations for use.

2.04 Use of Assessment in General and with Special Populations

(a) Psychologists who perform interventions or administer, score, interpret, or use assessment techniques are familiar with the reliability, validation, and related standardization or outcome studies of, and proper applications and uses of, the techniques they use.

(b) Psychologists recognize limits to the certainty with which diagnoses, judgments, or predictions can be made about individuals.

(c) Psychologists attempt to identify situations in which particular interventions or assessment techniques or norms may not be applicable or may require adjustment in administration or interpretation because of factors such as individuals' gender, age, race, ethnicity, national origin, religion, sexual orientation, disability, language, or socioeconomic status.

2.05 Interpreting Assessment Results

When interpreting assessment results, including automated interpretations, psychologists take into account the various test factors and characteristics of the person being assessed that might affect psychologists' judgments or reduce the accuracy of their interpretations. They indicate any significant reservations they have about the accuracy or limitations of their interpretations.

2.06 Unqualified Persons

Psychologists do not promote the use of psychological assessment techniques by unqualified persons. (See also Standard 1.22, Delegation to and Supervision of Subordinates.)

2.07 Obsolete Tests and Outdated Test Results

(a) Psychologists do not base their assessment or intervention decisions or recommendations on data or test results that are outdated for the current purpose.

(b) Similarly, psychologists do not base such decisions or recommendations on tests and measures that are obsolete and not useful for the current purpose.

2.08 Test Scoring and Interpretation Services

(a) Psychologists who offer assessment or scoring procedures to other professionals accurately describe the purpose, norms, validity, reliability, and applications of the procedures and any special qualifications applicable to their use.

(b) Psychologists select scoring and interpretation services (including automated services) on the basis of evidence of the validity of the program and procedures as well as on other appropriate considerations.

(c) Psychologists retain appropriate responsibility for the appropriate application, interpretation, and use of assessment instruments, whether they score and interpret such tests themselves or use automated or other services.

2.09 Explaining Assessment Results

Unless the nature of the relationship is clearly explained to the person being assessed in advance and precludes provision of an explanation of results (such as in some organizational consulting, preemployment or security screenings, and forensic evaluations), psychologists ensure that an explanation of the results is provided using language that is reasonably understandable to the person assessed or to another legally authorized person on behalf of the client. Regardless of whether the scoring and interpretation are done by the psychologist, by assistants, or by automated or other outside services, psychologists take reasonable steps to ensure that appropriate explanations of results are given.

2.10 Maintaining Test Security

Psychologists make reasonable efforts to maintain the integrity and security of tests and other assessment techniques consistent with law, contractual obligations, and in a manner that permits compliance with the requirements of this Ethics Code. (See also Standard 1.02, Relationship of Ethics and Law.)

3. ADVERTISING AND OTHER PUBLIC STATEMENTS

3.01 Definition of Public Statements

Psychologists comply with this Ethics Code in public statements relating to their professional services, products, or publications or to the field of psychology. Public statements include but are not limited to paid or unpaid advertising, brochures, printed matter, directory listings, personal resumes or curricula vitae, interviews or comments for use in media, statements in legal proceedings, lectures and public oral presentations, and published materials.

3.02 Statements by Others

(a) Psychologists who engage others to create or place public statements that promote their professional practice, products, or activities retain professional responsibility for such statements.

(b) In addition, psychologists make reasonable efforts to prevent others whom they do not control (such as employers, publishers, sponsors, organizational clients, and representatives of the print or broadcast media) from making deceptive statements concerning psychologists' practice or professional or scientific activities.

(c) If psychologists learn of deceptive statements about their work made by others, psychologists make reasonable efforts to correct such statements.

(d) Psychologists do not compensate employees of press, radio, television, or other communication media in return for publicity in a news item.

(e) A paid advertisement relating to the psychologist's activities must be identified as such, unless it is already apparent from the context.

3.03 Avoidance of False or Deceptive Statements

(a) Psychologists do not make public statements that are false, deceptive, misleading, or fraudulent, either because of what they state, convey, or suggest or because of what they omit, concerning their research, practice, or other work activities or those of persons or organizations with which they are affiliated. As examples (and not in limitation) of this standard, psychologists do not make false or deceptive statements concerning (1) their training, experience, or competence; (2) their academic degrees; (3) their credentials; (4) their institutional or association affiliations; (5) their services; (6) the scientific or clinical basis for, or results or degree of success of, their services; (7) their fees; or (8) their publications or research findings. (See also Standards 6.15, Deception in Research, and 6.18, Providing Participants with Information about the Study.)

(b) Psychologists claim as credentials for their psychological work, only degrees that (1) were earned from a regionally accredited educational institution or (2) were the basis for psychology licensure by the state in which they practice.

3.04 Media Presentations

When psychologists provide advice or comment by means of public lectures, demonstrations, radio or television programs, prerecorded tapes, printed articles, mailed material, or other media, they take reasonable precautions to ensure that (1) the statements are based on appropriate psychological literature and practice, (2) the statements are otherwise consistent with this Ethics Code, and (3) the recipients of the information are not encouraged to infer that a relationship has been established with them personally.

3.05 Testimonials

Psychologists do not solicit testimonials from current psychotherapy clients or patients or other persons who because of their particular circumstances are vulnerable to undue influence.

3.06 In-person Solicitation

Psychologists do not engage, directly or through agents, in uninvited in-person solicitation of business from actual or potential psychotherapy patients or clients or other persons who because of their particular circumstances are vulnerable to undue influence. However, this does not preclude attempting to implement appropriate collateral contacts with significant others for the purpose of benefiting an already engaged therapy patient.

4. THERAPY

4.1 Structuring the Relationship

(a) Psychologists discuss with clients or patients as early as is feasible in the therapeutic relationship appropriate issues, such as the nature and anticipated course of therapy, fees, and confidentiality. (See also Standards 1.25, Fees and Financial Arrangements, and 5.01, Discussing the Limits of Confidentiality.)

(b) When the psychologist's work with clients or patients will be supervised, the above discussion includes that fact, and the name of the supervisor, when the supervisor has legal responsibility for the case.

(c) When the therapist is a student intern, the client or patient is informed of that fact.

(d) Psychologists make reasonable efforts to answer patients' questions and to avoid apparent misunderstandings about therapy. Whenever possible, psychologists provide oral and/or written information, using language that is reasonably understandable to the patient or client.

4.02 Informed Consent to Therapy

(a) Psychologists obtain appropriate informed consent to therapy or related procedures, using language that is reasonably understandable to participants. The content of informed consent will vary depending on many circumstances; however, informed consent generally implies that the person (1) has the capacity to consent, (2) has been informed of significant information concerning the procedure, (3) has freely and without undue influence expressed consent, and (4) consent has been appropriately documented.

(b) When persons are legally incapable of giving informed consent, psychologists obtain informed permission from a legally authorized person, if such substitute consent is permitted by law.

(c) In addition, psychologists (1) inform those persons who are legally incapable of giving informed consent about the proposed interventions in a manner commensurate with the persons' psychological capacities, (2) seek their assent to those interventions, and (3) consider such persons' preferences and best interests.

4.03 Couple and Family Relationships

(a) When a psychologist agrees to provide services to several persons who have a relationship (such as husband and wife or parents and children), the psychologist attempts to clarify at the outset (1) which of the individuals are patients or clients and (2) the relationship the psychologist will have with each person. This clarification includes the role of the psychologist and the probable uses of the services provided or the information obtained. (See also Standard 5.01, Discussing the Limits of Confidentiality.)

(b) As soon as it becomes apparent that the psychologist may be called on to perform potentially conflicting roles (such as marital counselor to husband and wife, and then witness for one party in a divorce proceeding), the psychologist attempts to clarify and adjust, or withdraw from, roles appropriately. (See also Standard 7.03, Clarification of Role, under Forensic Activities.)

4.04 Providing Mental Health Services to Those Served by Others

In deciding whether to offer or provide services to those already receiving mental health services elsewhere, psychologists carefully consider the treatment issues and the potential patient's or client's welfare. The psychologist discusses these issues with the patient or client, or another legally authorized person on behalf of the client, in order to minimize the risk of confusion and conflict, consults with the other service providers when appropriate, and proceeds with caution and sensitivity to the therapeutic issues.

4.05 Sexual Intimacies with Current Patients or Clients

Psychologists do not engage in sexual intimacies with current patients or clients.

4.06 Therapy with Former Sexual Partners

Psychologists do not accept as therapy patients or clients persons with whom they have engaged in sexual intimacies.

4.07 Sexual Intimacies with Former Therapy Patients

(a) Psychologists do not engage in sexual intimacies with a former therapy patient or client for at least two years after cessation or termination of professional services.

(b) Because sexual intimacies with a former therapy patient or client are so frequently harmful to the patient or client, and because such intimacies undermine public confidence in the psychology profession and thereby deter the public's use of needed services, psychologists do not engage in sexual intimacies with former therapy patients and clients even after a two-year interval except in the most unusual circumstances. The psychologist who engages in such activity after the two years following cessation or termination of treatment bears the burden of demonstrating that there has been no exploitation, in light of all relevant factors, including (1) the amount of time that has passed since therapy terminated, (2) the nature and duration of the therapy, (3) the circumstances of termination, (4) the patient's or client's personal history, (5) the patient's or client's current mental status, (6) the likelihood of adverse impact on the patient or client and others, and (7) any statements or actions made by the therapist during the course of therapy suggesting or inviting the possibility of a posttermination sexual or romantic relationship with the patient or client. (See also Standard 1.17, Multiple Relationships.)

4.08 Interruption of Services

(a) Psychologists make reasonable efforts to plan for facilitating care in the event that psychological services are interrupted by factors such as the psychologist's illness, death, unavailability, or relocation or by the client's relocation or financial limitations. (See also Standard 5.09, Preserving Records and Data.)

(b) When entering into employment or contractual relationships, psychologists provide for orderly and appropriate resolution of responsibility for patient or client care in the event that the employment or contractual relationship ends, with paramount consideration given to the welfare of the patient or client.

4.09 Terminating the Professional Relationship

(a) Psychologists do not abandon patients or clients. (See also Standard 1.25e, under Fees and Financial Arrangements.)

(b) Psychologists terminate a professional relationship when it becomes reasonably clear that the patient or client no longer needs the service, is not benefiting, or is being harmed by continued service.

(c) Prior to termination for whatever reason, except where precluded by the patient's or client's conduct, the psychologist discusses the patient's or client's views and needs, provides appropriate pretermination counseling, suggests alternative service providers as appropriate, and takes other reasonable steps to facilitate transfer of responsibility to another provider if the patient or client needs one immediately.

5. PRIVACY AND CONFIDENTIALITY

These standards are potentially applicable to the professional and scientific activities of all psychologists.

5.01 Discussing the Limits of Confidentiality

(a) Psychologists discuss with persons and organizations with whom they establish a scientific or professional relationship (including, to the extent feasible, minors and their legal representatives) (1) the relevant limitations on confidentiality, including limitations where applicable in group, marital, and family therapy or in organizational consulting, and (2) the foreseeable uses of the information generated through their services.

(b) Unless it is not feasible or is contraindicated, the discussion of confidentiality occurs at the outset of the relationship and thereafter as new circumstances may warrant.

(c) Permission for electronic recording of interviews is secured from clients and patients.

5.02 Maintaining Confidentiality

Psychologists have a primary obligation and take reasonable precautions to respect the confidentiality rights of those with whom they work or consult, recognizing that confidentiality may be established by law, institutional rules, or professional or scientific relationships. (See also Standard 6.26, Professional Reviewers.)

5.03 Minimizing Intrusions on Privacy

(a) In order to minimize intrusions on privacy, psychologists include in written and oral reports, consultations, and the like, only information germane to the purpose for which the communication is made.

(b) Psychologists discuss confidential information obtained in clinical or consulting relationships, or evaluative data concerning patients, individual or organizational clients, students, research participants, supervisees, and employees, only for appropriate scientific or professional purposes and only with persons clearly concerned with such matters.

5.04 Maintenance of Records

Psychologists maintain appropriate confidentiality in creating, storing, accessing, transferring, and disposing of records under their control, whether these are written, automated, or in any other medium. Psychologists maintain and dispose of records in accordance with law and in a manner that permits compliance with the requirements of this Ethics Code.

5.05 Disclosures

(a) Psychologists disclose confidential information without the consent of the individual only as mandated by law, or where permitted by law for a valid purpose, such as (1) to provide needed professional services to the patient or the individual or organizational client, (2) to obtain appropriate professional consultations, (3) to protect the patient or client or others from harm, or (4) to obtain payment for services, in which instance disclosure is limited to the minimum that is necessary to achieve the purpose.

(b) Psychologists also may disclose confidential information with the appropriate consent of the patient or the individual or organizational client (or of another legally authorized person on behalf of the patient or client), unless prohibited by law.

5.06 Consultations

When consulting with colleagues, (1) psychologists do not share confidential information that reasonably could lead to the identification of a patient, client, research participant, or other person or organization with whom they have a confidential relationship unless they have obtained the prior consent of the person or organization or the disclosure cannot be avoided, and (2) they share information only to the extent necessary to achieve the purposes of the consultation. (See also Standard 5.02, Maintaining Confidentiality.)

5.07 Confidential Information in Databases

(a) If confidential information concerning recipients of psychological services is to be entered into databases or systems of records available to persons whose access has not been consented to by the recipient, then psychologists use coding or other techniques to avoid the inclusion of personal identifiers.

(b) If a research protocol approved by an institutional review board or similar body requires the inclusion of personal identifiers, such identifiers are deleted before the information is made accessible to persons other than those of whom the subject was advised.

(c) If such deletion is not feasible, then before psychologists transfer such data to others or review such data collected by others, they take reasonable steps to determine that appropriate consent of personally identifiable individuals has been obtained.

5.08 Use of Confidential Information for Didactic or Other Purposes

(a) Psychologists do not disclose in their writings, lectures, or other public media, confidential, personally identifiable information concerning their patients, individual or organizational clients, students, research participants, or other recipients of their services that they obtained during the course of their work, unless the person or organization has consented in writing or unless there is other ethical or legal authorization for doing so.

(b) Ordinarily, in such scientific and professional presentations, psychologists disguise confidential information concerning such persons or organizations so that they are not individually identifiable to others and so that discussions do not cause harm to subjects who might identify themselves.

5.09 Preserving Records and Data

A psychologist makes plans in advance so that confidentiality of records and data is protected in the event of the psychologist's death, incapacity, or withdrawal from the position or practice.

5.10 Ownership of Records and Data

Recognizing that ownership of records and data is governed by legal principles, psychologists take reasonable and lawful steps so that records and data remain available to the extent needed to serve the best interests of patients, individual or organizational clients, research participants, or appropriate others.

5.11 Withholding Records for Nonpayment

Psychologists may not withhold records under their control that are requested and imminently needed for a patient's or client's treatment solely because payment has not been received, except as otherwise provided by law.

6. TEACHING, TRAINING SUPERVISION, RESEARCH, AND PUBLISHING

6.01 Design of Education and Training Programs

Psychologists who are responsible for education and training programs seek to ensure that the programs are competently designed, provide the proper experiences, and meet the requirements for licensure, certification, or other goals for which claims are made by the program.

6.02 Descriptions of Education and Training Programs

(a) Psychologists responsible for education and training programs seek to ensure that there is a current and accurate description of the program content, training goals and objectives, and requirements that must be met for satisfactory completion of the program. This information must be made readily available to all interested parties.

(b) Psychologists seek to ensure that statements concerning their course outlines are accurate and not misleading, particularly regarding the subject matter to be covered, bases for evaluating progress, and the nature of course experiences. (See also Standard 3.03, Avoidance of False or Deceptive Statements.)

(c) To the degree to which they exercise control, psychologists responsible for announcements, catalogs, brochures, or advertisements describing workshops, seminars, or other non-degree-granting educational programs ensure that they accurately describe the audience for which the program is intended, the educational objectives, the presenters, and the fees involved.

6.03 Accuracy and Objectivity in Teaching

(a) When engaged in teaching or training, psychologists present psychological information accurately and with a reasonable degree of objectivity.

(b) When engaged in teaching or training, psychologists recognize the power they hold over students or supervisees and therefore make reasonable efforts to avoid engaging in conduct that is personally demeaning to students or supervisees. (See also Standards 1.09, Respecting Others, and 1.12, Other Harassment.)

6.04 Limitation on Teaching

Psychologists do not teach the use of techniques or procedures that require specialized training, licensure, or expertise, including but not limited to hypnosis, biofeedback, and projective techniques, to individuals who lack the prerequisite training, legal scope of practice, or expertise.

6.05 Assessing Student and Supervisee Performance

(a) In academic and supervisory relationships, psychologists establish an appropriate process for providing feedback to students and supervisees.

(b) Psychologists evaluate students and supervisees on the basis of their actual performance on relevant and established program requirements.

6.06 Planning Research

(a) Psychologists design, conduct, and report research in accordance with recognized standards of scientific competence and ethical research.

(b) Psychologists plan their research so as to minimize the possibility that results will be misleading.

(c) In planning research, psychologists consider its ethical acceptability under the Ethics Code. If an ethical issue is unclear, psychologists seek to resolve the issue through consultation with institutional review boards, animal care and use committees, peer consultations, or other proper mechanisms.

(d) Psychologists take reasonable steps to implement appropriate protections for the rights and welfare of human participants, other persons affected by the research, and the welfare of animal subjects.

6.07 Responsibility

(a) Psychologists conduct research competently and with due concern for the dignity and welfare of the participants.

(b) Psychologists are responsible for the ethical conduct of research conducted by them or by others under their supervision or control.

(c) Researchers and assistants are permitted to perform only those tasks for which they are appropriately trained and prepared.

(d) As part of the process of development and implementation of research projects, psychologists consult those with expertise concerning any special population under investigation or most likely to be affected.

6.08 Compliance with Law and Standards

Psychologists plan and conduct research in a manner consistent with federal and state law and regulations, as well as professional standards governing the conduct of research, and particularly those standards governing research with human participants and animal subjects.

6.09 Institutional Approval

Psychologists obtain from host institutions or organizations appropriate approval prior to conducting research, and they provide accurate information about their research proposals. They conduct the research in accordance with the approved research protocol.

6.10 Research Responsibilities

Prior to conducting research (except research involving only anonymous surveys, naturalistic observations, or similar research), psychologists enter into an agreement with participants that clarifies the nature of the research and the responsibilities of each party.

6.11 Informed Consent to Research

(a) Psychologists use language that is reasonably understandable to research participants in obtaining their appropriate informed consent (except as provided in Standard 6.12, Dispensing with Informed Consent). Such informed consent is appropriately documented.

(b) Using language that is reasonably understandable to participants, psychologists inform participants of the nature of the research; they inform participants that they are free to participate or to decline to participate or to withdraw from the research; they explain the foreseeable consequences of declining or withdrawing; they inform participants of significant factors that may be expected to influence their willingness to participate (such as risks, discomfort, adverse effects, or limitations on confidentiality, except as provided in Standard 6.15, Deception in Research); and they explain other aspects about which the prospective participants inquire.

(c) When psychologists conduct research with individuals such as students or subordinates, psychologists take special care to protect the prospective participants from adverse consequences of declining or withdrawing from participation.

(d) When research participation is a course requirement or opportunity for extra credit, the prospective participant is given the choice of equitable alternative activities.

(e) For persons who are legally incapable of giving informed consent, psychologists nevertheless (1) provide an appropriate explanation, (2) obtain the participant's assent, and (3) obtain appropriate permission from a legally authorized person, if such substitute consent is permitted by law.

6.12 Dispensing with Informed Consent

Before determining that planned research (such as research involving only anonymous questionnaires, naturalistic observations, or certain kinds of archival research) does not require the informed consent of research participants, psychologists consider applicable regulations and institutional review board requirements, and they consult with colleagues as appropriate.

6.13 Informed Consent in Research Filming or Recording

Psychologists obtain informed consent from research participants prior to filming or recording them in any form, unless the research involves simply naturalistic observations in public places and it is not anticipated that the recording will be used in a manner that could cause personal identification or harm.

6.14 Offering Inducements for Research Participants

(a) In offering professional services as an inducement to obtain research participants, psychologists make clear the nature of the services, as well as the risks, obligations, and limitations. (See also Standard 1.18, Barter [with Patients or Clients].)

(b) Psychologists do not offer excessive or inappropriate financial or other inducements to obtain research participants, particularly when it might tend to coerce participation.

6.15 Deception in Research

(a) Psychologists do not conduct a study involving deception unless they have determined that the use of deceptive techniques is justified by the study's prospective scientific, educational, or applied value and that equally effective alternative procedures that do not use deception are not feasible.

(b) Psychologists never deceive research participants about significant aspects that would affect their willingness to participate, such as physical risks, discomfort, or unpleasant emotional experiences.

(c) Any other deception that is an integral feature of the design and conduct of an experiment must be explained to participants as early as is feasible, preferably at the conclusion of their participation, but no later than at the conclusion of the research. (See also Standard 6.18, Providing Participants with Information about the Study.)

6.16 Sharing and Utilizing Data

Psychologists inform research participants of their anticipated sharing or further use of personally identifiable research data and of the possibility of unanticipated future uses.

6.17 Minimizing Invasiveness

In conducting research, psychologists interfere with the participants or milieu from which data are collected only in a manner that is warranted by an appropriate research design and that is consistent with psychologists' roles as scientific investigators.

6.18 Providing Participants with Information about the Study

(a) Psychologists provide a prompt opportunity for participants to obtain appropriate information about the nature, results, and conclusions of the research, and psychologists attempt to correct any misconceptions that participants may have.

(b) If scientific or humane values justify delaying or withholding this information, psychologists take reasonable measures to reduce the risk of harm.

6.19 Honoring Commitments

Psychologists take reasonable measures to honor all commitments they have made to research participants.

6.20 Care and Use of Animals in Research

(a) Psychologists who conduct research involving animals treat them humanely.

(b) Psychologists acquire, care for, use, and dispose of animals in compliance with current federal, state, and local laws and regulations, and with professional standards.

(c) Psychologists trained in research methods and experienced in the care of laboratory animals supervise all procedures involving animals and are responsible for ensuring appropriate consideration of their comfort, health, and humane treatment.

(d) Psychologists ensure that all individuals using animals under their supervision have received instruction in research methods and in the care, maintenance, and handling of the species being used, to the extent appropriate to their role.

(e) Responsibilities and activities of individuals assisting in a research project are consistent with their respective competencies.

(f) Psychologists make reasonable efforts to minimize the discomfort, infection, illness, and pain of animal subjects.

(g) A procedure subjecting animals to pain, stress, or privation is used only when an alternative procedure is unavailable and the goal is justified by its prospective scientific, educational, or applied value.

(h) Surgical procedures are performed under appropriate anesthesia; techniques to avoid infection and minimize pain are followed during and after surgery.

(i) When it is appropriate that the animal's life be terminated, it is done rapidly, with an effort to minimize pain, and in accordance with accepted procedures.

6.21 Reporting of Results

(a) Psychologists do not fabricate data or falsify results in their publications.

(b) If psychologists discover significant errors in their published data, they take reasonable steps to correct such errors in a correction, retraction, erratum, or other appropriate publication means.

6.22 Plagiarism

Psychologists do not present substantial portions or elements of another's work or data as their own, even if the other work or data source is cited occasionally.

6.23 Publication Credit

(a) Psychologists take responsibility and credit, including authorship credit, only for work they have actually performed or to which they have contributed.

(b) Principal authorship and other publication credits accurately reflect the relative scientific or professional contributions of the individuals involved, regardless of their relative status. Mere possession of an institutional position, such as Department Chair, does not justify authorship credit. Minor contributions to the research or to the writing for publications are appropriately acknowledged, such as in footnotes or in an introductory statement.

(c) A student is usually listed as principal author on any multiple-authored article that is substantially based on the student's dissertation or thesis.

6.24 Duplicate Publication of Data

Psychologists do not publish, as original data, data that have been previously published. This does not preclude republishing data when they are accompanied by proper acknowledgment.

6.25 Sharing Data

After research results are published, psychologists do not withhold the data on which their conclusions are based from other competent professionals who seek to verify the substantive claims through reanalysis and who intend to use such data only for that purpose, provided that the confidentiality of the participants can be protected and unless legal rights concerning proprietary data preclude their release.

6.26 Professional Reviewers

Psychologists who review material submitted for publication, grant, or other research proposal review respect the confidentiality of and the proprietary rights in such information of those who submitted it.

7. FORENSIC ACTIVITIES

7.01 Professionalism

Psychologists who perform forensic functions, such as assessments, interviews, consultations, reports, or expert testimony, must comply with all other provisions of this Ethics Code to the extent that they apply to such activities. In addition, psychologists base their forensic work on appropriate knowledge of and competence in the areas underlying such work, including specialized knowledge concerning special populations. (See also Standards 1.06, Basis for Scientific and Professional Judgments; 1.08, Human Differences; 1.15, Misuse of Psychologists' Influence; and 1.23, Documentation of Professional and Scientific Work.)

7.02 Forensic Assessments

(a) Psychologists' forensic assessments, recommendations, and reports are based on information and techniques (including personal interviews of the individual, when appropriate) sufficient to provide appropriate substantiation for their findings. (See also Standards 1.03, Professional and Scientific Relationship; 1.23, Documentation of Professional and Scientific Work; 2.01, Evaluation, Diagnosis, and Interventions in Professional Context; and 2.05, Interpreting Assessment Results.)

(b) Except as noted in (c), below, psychologists provide written or oral forensic reports or testimony of the psychological characteristics of an individual only after they have conducted an examination of the individual adequate to support their statements or conclusions.

(c) When, despite reasonable efforts, such an examination is not feasible, psychologists clarify the impact of their limited information on the reliability and validity of their reports and testimony, and they appropriately limit the nature and extent of their conclusions or recommendations.

7.03 Clarification of Role

In most circumstances, psychologists avoid performing multiple and potentially conflicting roles in forensic matters. When psychologists may be called on to serve in more than one role in a legal proceeding—for example, as consultant or expert for one party or for the court and as a fact witness—they clarify role expectations and the extent of confidentiality in advance to the extent feasible, and thereafter as changes occur, in order to avoid compromising their professional judgment and objectivity and in order to avoid misleading others regarding their role.

7.04 Truthfulness and Candor

(a) In forensic testimony and reports, psychologists testify truthfully, honestly, and candidly and, consistent with applicable legal procedures, describe fairly the bases for their testimony and conclusions.

(b) Whenever necessary to avoid misleading, psychologists acknowledge the limits of their data or conclusions.

7.05 Prior Relationships

A prior professional relationship with a party does not preclude psychologists from testifying as fact witnesses or from testifying to their services to the extent permitted by applicable law. Psychologists appropriately take into account ways in which the prior relationship might affect their professional objectivity or opinions and disclose the potential conflict to the relevant parties.

7.06 Compliance with Law and Rules

In performing forensic roles, psychologists are reasonably familiar with the rules governing their roles. Psychologists are aware of the occasionally competing demands placed upon them by these principles and the requirements of the court system, and attempt to resolve these conflicts by making known their commitment to this Ethics Code and taking steps to resolve the conflict in a responsible manner. (See also Standard 1.02, Relationship of Ethics and Law.)

8. RESOLVING ETHICAL ISSUES

8.01 Familiarity with Ethics Code

Psychologists have an obligation to be familiar with this Ethics Code, other applicable ethics codes, and their application to psychologists' work. Lack of awareness or misunderstanding of an ethical standard is not itself a defense to a charge of unethical conduct.

8.02 Confronting Ethical Issues

When a psychologist is uncertain whether a particular situation or course of action would violate this Ethics Code, the psychologist ordinarily consults with other psychologists knowledgeable about ethical issues, with state or national psychology ethics committees, or with other appropriate authorities in order to choose a proper response.

8.03 Conflicts between Ethics and Organizational Demands

If the demands of an organization with which psychologists are affiliated conflict with this Ethics Code, psychologists clarify the nature of the conflict, make known their commitment to the Ethics Code, and to the extent feasible, seek to resolve the conflict in a way that permits the fullest adherence to the Ethics Code.

8.04 Informal Resolution of Ethical Violations

When psychologists believe that there may have been an ethical violation by another psychologist, they attempt to resolve the issue by bringing it to the attention of that individual if an informal resolution appears appropriate and the intervention does not violate any confidentiality rights that may be involved.

8.05 Reporting Ethical Violations

If an apparent ethical violation is not appropriate for informal resolution under Standard 8.04 or is not resolved properly in that fashion, psychologists take further action appropriate to the situation, unless such action conflicts with confidentiality rights in ways that cannot be resolved. Such action might include referral to state or national committees on professional ethics or to state licensing boards.

8.06 Cooperating with Ethics Committees

Psychologists cooperate in ethics investigations, proceedings, and resulting requirements of the APA or any affiliated state psychological association to which they belong. In doing so, they make reasonable efforts to resolve any issues as to confidentiality. Failure to cooperate is itself an ethics violation.

8.07 Improper Complaints

Psychologists do not file or encourage the filing of ethics complaints that are frivolous and are intended to harm the respondent rather than to protect the public.

APPENDIX C

OUTLINE FOR SESSION WRITE-UP

Conducting counseling sessions with the support and feedback of a supervisor is an important part of a trainee's professional development. In most counselor education programs, sessions are recorded on either audio or video tape. Once a session is completed, the trainee plays back the recording, providing himself or herself with feedback that can lead to ideas for future sessions and recognition of areas in which improvement is possible. Advanced counselors as well as trainees listen to or watch such tapes to develop a full picture of what happened in a session: important client emotions, beliefs, and personality characteristics not fully perceived during the session, significant moments or turning points, and notable things the counselor did or did not do. Receiving organized feedback from a supervisor and giving organized feedback to oneself are major vehicles both for improving one's work with a given client and for building one's professional skills.

The preparation of written critiques of one's counseling sessions is an integral part of the supervisory process. Each critique should show the counselor's thoughtful consideration of the progress of the counseling process with special attention to (1) the client's progress toward the resolution of counseling concerns, and (2) the counselor's progress as the planner and facilitator of the counseling.

A critique is similar to a case study in that it shows evidence of careful observation of a client and thought about the meanings of his or her behaviors. Rationales are needed when inferences are stated. A critique differs from a case study in that it is also a process report in which the counselor actively seeks formative feedback from the supervisor.

The following outline is offered as an aid to producing a written analysis of a counseling session, especially one that has been recorded. Used carefully, the outline can help a counselor organize thoughts about the client, the session, and the quality of his or her work. A copy of the write-up will provide the supervisor with a quick yet intensive picture of the session and thus help him or her develop feedback that will be of greatest possible utility to both counselor and client.

The outline contains six major sections, whose headings are given below, along with indications of the kind of content that each section should contain. The content described for each category is suggestive of things that should be included, but each question need not be answered for each session and additional material ought to be included as appropriate. In continuing counseling, the critiques should be regarded as a series, and there is no need to repeat detail that has already been presented. Bridges from one critique to another that show new developments are helpful. The last section of the outline—headed "Help!"—reflects our experience as supervisors that a student's receptivity to feedback is directly related to his or her ability to clarify the kind of feedback that would be most helpful.

Background Information

Why did the client approach you (or why did you approach the client)? What were the important concerns and circumstances that brought the two of you together? If third parties were involved, what were the observations and concerns of the third party? State demographic information (such as age, grade in school, employment, family unit, and history) that seems relevant to the presenting problem.

Overview of the Session

What did you talk about? What were the dominant issues and themes for this session? If the session was not the initial meeting, what were your process and outcome goals going into the session? Did you make progress toward these goals? Did you find it necessary to do something different because of the client's priorities? (This section should be fairly brief, leaving details of significant interactions for the following section.)

Observations and Diagnostic Assessment

What observations and impressions do you have about your client and his or her life space? How intractable are the barriers to growth? How strong are the coping skills? What is the etiology of the client's present psychological capacity or incapacity? What is he or she trying to accomplish by various behaviors? What are the influences of significant others in the client's life? How effectively does the client handle relationships with significant others? What are your hypotheses about the client that may serve to form your counseling interventions? (DSM III-R diagnostic categories may be used if you believe they are appropriate. Chapter 7 provides a structure for addressing many of these questions.)

Observations about Self

Describe significant themes and patterns you observed in your own behavior, noting what you did that you considered especially effective and areas that were troublesome for you. Describe your own internal experiencing during the session, with special focus on times or places where you felt confused, tense, disappointed, annoyed, disapproving, or controlling. Pay attention to any tendency to do for the client what the client needs to do for himself or herself.

Plans for the Next Session

How do you hope to follow up in subsequent sessions? What issues and concerns do you think will be worthwhile to explore? What process goals will you try to accomplish?

Help!

Specify what kind of help you would like, either from your practicum supervisor or from fellow students, about this client, this session, and your helping efforts.

AUTHOR INDEX

Subject Index

TO THE OWNER OF THIS BOOK:

We hope that you have found *The Counseling Process*, Fourth Edition, useful. So that this book can be improved in a future edition, would you take the time to complete this sheet and return it? Thank you.

School and address: ______________________________

Department: ______________________________

Instructor's name: ______________________________

1. What I like most about this book is: ______________________________

____

2. What I like least about this book is: ______________________________

____

3. My general reaction to this book is: ______________________________

____

4. The name of the course in which I used this book is: ______________________________

____

5. Were all of the chapters of the book assigned for you to read? ______________________________

   If not, which ones weren't? ______________________________

____

6. In the space below, or on a separate sheet of paper, please write specific suggestions for improving this book and anything else you'd care to share about your experience in using the book.

____

____

____

____

____

____

Optional:

Your name: ______________________________ Date: ______________________

May Brooks/Cole quote you either in promotion for *The Counseling Process*, Fourth Edition, or in future publishing ventures?

Yes: _______ No: _______

Sincerely,

Lewis E. Patterson
Elizabeth Reynolds Welfel

FOLD HERE

NO POSTAGE NECESSARY IF MAILED IN THE UNITED STATES

BUSINESS REPLY MAIL
FIRST CLASS PERMIT NO. 358 PACIFIC GROVE, CA

POSTAGE WILL BE PAID BY ADDRESSEE

ATT: *Lewis E. Patterson & Elizabeth Reynolds Welfel*

Brooks/Cole Publishing Company
511 Forest Lodge Road
Pacific Grove, California 93950-9968

FOLD HERE

Brooks/Cole is dedicated to publishing quality publications for education in the human services fields. If you are interested in learning more about our publications, please fill in your name and address and request our latest catalogue, using this prepaid mailer.

Name: ______________________________

Street Address: ______________________________

City, State, and Zip: ______________________________

FOLD HERE

BUSINESS REPLY MAIL

FIRST CLASS PERMIT NO. 358 PACIFIC GROVE, CA

POSTAGE WILL BE PAID BY ADDRESSEE

ATT: *Human Services Catalogue*

Brooks/Cole Publishing Company
511 Forest Lodge Road
Pacific Grove, California 93950-9968

FOLD HERE

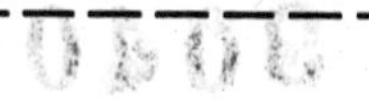